ARTHUR RANDELL
FENMAN

Bygones presented by Anglia Television

ARTHUR RANDELL
FENMAN

talking to
Dick Joice

The Boydell Press · Ipswich

First published by The Boydell Press Ltd
PO Box 24 Ipswich IP1 1JJ
Reprinted 1976
Reprinted 1977

ISBN 0 85115 076 4

Printed in Great Britain by
Lowe & Brydone Printers Limited, Thetford, Norfolk

FOREWORD

This book is about Arthur Randell – Fenman. I first met him about 20 years ago when I interviewed him for a television programme about catching moles.

Since then we have talked or rather I have listened, occasionally in the Studio, but mostly with our feet under a table or just sitting. Arthur calls his stories 'his yarns' – he usually starts by saying 'did I ever tell you . . .' and finishes with 'did you ever hear such squit'. They don't begin in any particular place and often don't end, but they leave a vivid picture in the mind of the listener of whatever aspect of his life Arthur is talking about.

In order to capture some of the enthusiasm with which Arthur tells his stories, a few words in each chapter are spelt phonetically which also conveys a little of the Fen dialect.

By some standards Arthur hasn't had a very spectacular life, indeed he hasn't moved very far in his 75 years from where he was born near Wisbech. Three generations of Randell's lived in that area before Arthur came along in 1901 – they were all molecatchers. He has often told me that Fenmen are different. I've never really discovered what he means by this – but he's a good example! For although he hasn't gained fame for any outstanding deed, if you mention his name in that flat fertile county everyone knows who you're talking about.

It's quite uncanny how Arthur can hold an audience by simply 'yarning on' about everyday events in his life. I hope in these few brief pages to have captured some of what 'he say'.

DICK JOICE

OUR HOME AND VILLAGE

I was born nearly opposite the Church at Wiggenhall St Mary Magdalen which is right agin the Banks of the Great Ouse River. Our house, where all 13 of us children was born, was also where my Father was born. It was quite a small house (2 Bedrooms and 1 Living Room and a little old Kitchen). It was tiled and spouts carried the rain water into a big old Tank and that was our drinken and bathen water. During a long dry spell water was fetched from the River Ouse, allest when the Tide was running down, as that would be fresh water but if it was running up the water would be Salt or "Brackish". Sometimes a fella would come from Tottenhill where there was a Spring and he had a Hoss and Water-Cart and the water he sold at 1d a pailfull. As he could not get over Old Magdalen Bridge people still had to fetch it with Pails and Yokes from his Cart but they was glad to git it and it was used for Drinking only. When I was about 8 or 9 years old piped water was laid on by the Wisbech Water Board coming from big Springs at Marham so then we wunt short arter that. We allest lived in the Living Room, a big leafed Table covered over by a bit of Oil Cloth stood in the middle so if the younger children upset anything on the Table it could be wiped off ever so easy. Father and Mother each had a big old wooden Arm Chair and Fathers big Mollcatching Jacket was allest hung over the back of his. The Baby allest set, when it was old enough, next to Mother, and anyone of us others, who perhaps had been "putting their parts on" would have to set next to Father so he could, as he allest used to say, "jest give them a tap". The floor was allest covered with Coker-Matting, with 2 or 3 Pegged Rugs near the fire. Mother and us children made about 3 Pegged Rugs every year with bits cut from Old Coats, Trousers, etc, and although it was nearly allest a dark material it would allest hev a Diamond in some other colour in the middle. The Furniture consisted of one Chair for each of us, a wooden Stool, of course the Table, a Couch, Fathers Old Bureau, a Corner Cupboard full of Drugs and Corination Mug, and a Wicker Cradle (jest behind Mothers Chair). Above the Mantle-Piece hung an Assagai (throwing spear) which my Aunt Sally had brought home with her from South Africa. On the Mantle-Piece was 2 China Dorgs (bought by Father afore I was born, when he sold a Spaniel Bitch for a lot of money), 2 China Figures that when touched would nod their hids for several minutes. And in the middle was a little House that had bin in the Randells family for 150 years. A leather Strap allest hung agin the Fire-Place that was used to correct us kids if we din't behave ourselves. A very big Picture allest hung opposite the Fire and to us kids it allest seemed a bit of a mystery. The Picture was of a very Dark Beautiful lady and that had bin handed down for 2 or 3 generations. Strangers comen into the room would nearly allest admire that Picture as it somehow seemed to be alive, the Eyes seemed to be staring right into yer. In the Bureau was Fathers shoe mending tools, Haircutting Sizzors, small Nuts and Bolts, Nails, etc, but in a small compartment, allest kept locked, was his Poison for Varmin killen. Nobody only him ever unlocked that compartment and if he was gitting some out to use the next day or mixing some up to go Stack Dressing for Mice, a pin could be heard if it dropped on the floor, we daren't even scratch our snout. In the little old Kitchen was a big Flour Bin (held 10 stone Flour it did), a Table, 2 Chairs and a Brick Oven built into the Wall. Thats where Mother done her cooken and Bakin Bread. She would make 8 loaves of Bread 3 times a week and when the oven was hot she would cook apple Pies, Pumpkin Pies or anything else that she wanted cooken. Over the Fire in the Living Room was a Hake on which she allest hung the Big Iron Kettle (held a gallen water), a Big Boiler, or the Frying Pan (Frying Pan allest hung over the Fire), as although we allest kept a Fire-guard round the Fire, Mother allest said children could reach through the Bars and pull things over and git scalded. Upstairs the 2 Bedrooms was med into 3 by a thick Curtain that halved the biggest room of the two. A few texts hung on the walls and one thing that I shall always remember was Fathers clothes Chest. It was allest called a Coffar but it was a lovely bit of work with lots of engravings. Seventy year ago a Doctor would have give him £10 and had another the same size med for him but Father would not let it go. Saturday mornings was the time to clean knives and shoes. (All types of boots was called shoes). The Knives was cleaned by working them backards and forards on a Knife-Board on which had been scraped some Bath-Brick. The Bath-Brick when bought was nearly as big as a house-brick and it was scraped with a knife and the powder let fall on the knife-board. To clean the Shoes we used *Blacking* which was bought in small slabs wrapped up in paper. I think thats where the saying must have come from "Spit and Polish" as it took a lot of polish to make the shoes shine. Another thing I had never seen afore I was about 9 or 10 was Shaving Sticks. Father allest had a bit of the babys scented soap in his Shaving Tin, but quite a lot used Sunlight washing soap, and Safty Razors had not bin seen round our part, all used the Cut-throat type. You may wonder how we existed, well as you can see we did, but it was not on the fancy food that we got. The food we did git was *Pure*, the Bread we ate was all made from best Flour and what meat we had, although not fancy joints, was good. A Bullocks Head boiled up in the Big Iron Saucepan with plenty of Onions, Carrots and Peas and 20 minutes afore it was cooked about 9 or 10 light dumplins put in made a dinner fit for a King. The meat would boil off the bones and there would allest be enough left over to give us kids our breakfast. Bread would be crumbled in basins and the liquid boiled up agin and then poured over the bread and we went off to School as warm as toast on the coldest of mornings. Other mornings it would be Bread and Milk (skimmed) or Salt Sop, med with crumbled Bread, a small bit of Butter (to give it a sparkle) Salt and Pepper and Boiling Water. Dinner-times we would hev Apple or Pumpkin Pie, sometimes Pancakes, a Slice of Salt Breast of Pork and plenty of Vegetables and Tea-times it might be Rabbit Pie or Pigs Liver Dumplin and

Vegetables but Mother allest med sure there was plenty of Gravey so we could "Dip our bread in it". Saturday night Tea would allest be Fish of some sort – Bloaters, White Herring, Mackeral or it might be Sprats or sometimes Cockles or Mussels. Boats used to come up the River from Kings Lynn and the Fishermen would hawk the Shell Fish all round the village. The Cockles or Mussels would be boiled and we would hev them hot for tea and any left over would be covered with Vinegar and we would hev them for Sunday night Tea. Us kids allest had plenty to eat even if it won't posh grub but one thing we never did hev in our house was Margerine. Somebody once told Father that Margerine was med with the skimmings off the boilers where they was boiling Hoss Flesh and him being no scholar, took it in, so we allest did hev best Farm Butter. Another thing he wont eat was Corned Beef so I should think somebody must have told him they found a Hoss Shoe inside a tin. We eat Sparrow Pies, Pigs-Bellys, Pluck Pee, and hev had a hedgehog a few times but not Margerine or Corned Beef. There was not much in the village as regards amusements so it was all work while it was light and Bed when it was dark. Plenty a times I hev had to walk over $1\frac{1}{2}$ miles down an old mud drove afore I went to school to feed Fathers pigs as he being a Mollcatcher etc. would hev to go in the opposite direction. Arter feeding the Pigs I would hev another mile to go to School and my boots would be all mud so I stood the chance of gitting into trouble with the School Master. Saturdays and Holidays I was allest Fathers mate, going everywhere with him Mollcatchen (he allest sed "Molls"), Rat-catchen or Rabbit-catchen. I was taught how to make and set Molltraps, set Snares, Rabbit Traps, use a Gun and all there was to know about Dogs and Ferrets. Perhaps Mother would tell us when we got home that the chimney wanted sweeping so arter we had had our dinner Father would cut some Thorns, which he would make into the right sized bunch, tie a long piece of rope to them and I would go up a long ladder with the rope and thorns to the top of the chimney. Father would then go inside the house and I would drop the rope down inside the chimney and arter I had eased the thorns into the top of the chimney, he would pull it down inside. When Mothers oven or copper chimney wanted sweeping we would wait until she had a fire burning, git a pipe-bowl full of Gunpowder, wrap it up in several thicknesses of paper, throw it on the fire and hold a broom Handle tight aginst the copper door and when it had gone off there was soot laying about everywhere, but the chimney was cleared. Also hanging on our living room wall was things that are hardley ever seen nowadays, Button Hooks and Bellows. All mens heavy shoes was med with tough leather which arter being in the wet would be as hard as boards when dryed out. All heavy shoes was laced up with leather laces and it was a job to git them tight so a button Hook was used. There was various sorts and shapes, the ones Father and my brothers and self used was what Father had med from old Gimlets so they was strong and had a good handle. Other types was much lighter and was used by females to do their Button Shoes up, there often being perhaps a dozen buttons on each as the tops of the shoes reached to the calf of the leg. Childrens leather Leggins was allest a job to do up without a

Button Hook. As we had very few Toys to play with I have amused myself hours at playing at Engines with 3 or 4 Button Hooks. I would make a shed on the table with Dominoes and hang the Button Hooks together and pretend I had an ingin and thrashen tackle. When the fire wont draw very well a few puffs with the Bellows would soon make it begin to start flaming. Also hanging on the Stairs wall was an old Copper Warming Pan, which with a few hot ashes inside would soon warm a bed up although when it was very cold weather Mother would put a couple of Bricks in the oven and when hot would wrap them in some old flannel and put them in the bed. Beside the Warming Pan hung 2 Cart Lamps that we used if out at night with the Pony and Cart. They burned candles and was allest kept nice and shining. It was not often Father was out arter dark but he had a contract with the Police to drive anyone who had bin arrested to Terrington Police Station so they was kept ready for the road. Many a time he has bin called out of bed to drive the Village Policeman and prisoner to Terrington Lockup, one night, or rather early morning he driv the Policeman and 2 men who had done a big robbery and all four was handcuffed together. I have heard him say that he really did expect trouble that time as both men was big built chaps and one of them had bin a Boxer, but the Policeman who was a very big man told them if there was going to be any rough stuff he would be the one who would finish it. He said to them "You dint expect to see me walk into that Railway Waiting Room did you" and the Boxer said "No we dint and if it had bin anybody else we should have done him", so Woodcock (the Policeman) said "Don't start or I shall give you wuss then I give you afore" and coming home he told Father that when he was a Policeman in Norwich these two had knocked 2 Policemen about and Woodcock came rushing up and they set about him but he knocked both of them down and they was pleased to be locked up out of his way. In our present day when we see Police cars rushing about and we remember how

years ago the Village Bobby had a Bike to ride round on and his Superinten-
dent would visit him about once a fortnight riding in a Hoss and Trap driven
by a Constable. Another big change is heavy traffic. Years ago quite a lot of
heavy traffic was carried by Water, on Rivers and Canals. Every week at
Magdalen we would see the old Steam Tugs "Olga" and "Nancy", pulling
about 6 or 8 heavy loaded Barges from Lynn to Cambridge, etc, and every
week there would be other strings of Barges pulled by 2 hosses (one on each
Bank). They allest tried to go with the Tide and I hev seen them waiting at
Magdalen Bridge for the Tide to change. It was something us kids looked out
for as the river banks hev Styles about 3 or 4 feet high and the hosses would
hev to jump them. Hosses that was well used to jumping would allest stop a
few seconds to enable to get some slack rope but I hev seen some jump with a
tight rope and it has pulled the hoss over backards. With the Tugs it was
much better, they was so much closer to the Barges and old Black Bob (the
Engineer) on the "Nancy" allest had plenty of Steam. Before the new Bridge
was built (1914) the "Nancy" or the "Olga" could not git under the old
wooden bridge at High Tide unless the Funnel was lowered and Bob would
come up out of the Engine Room with his wiper and lower the Funnel until
the Tug was past the Bridge. When we heard the Tugs a blowin their Sirens
we would race up to the Bridge to see them pass under and we was allest sure
of a wave from the crew. Another time we would race to the Bridge was when
someone shouted "Eager Tide". This was a wall of water something like the
"Bore" on the River Severn and it was ruled by the moon. Except for the
annual trip to Hunstanton when the Scholars of the Church Sunday School
and members of the 2 Chapels all went, there was not a lot of entertainment
but every now and agin a Punch and Judy Show might come and stand in the
Cock Public House Yard and on July 29th it was Magdalen Fair. It lasted a
week with its roundabouts, swings, stalls, etc., all lit up when it got dark by Oil
Lamps which was hung at various points. The Lamps had naked burners that
made a Fizzing noise when alight. The Fair people was all relatives, there was
the Nunns, Harts and Wyatts and I know that the old Mr Wyatt was a
religious man and his Vans would never be shifted on a Sunday. Then once a
marionette show stood in the "Cock" Yard and gave a performance every
night and on Saturday arternoons for children. They was there 3 weeks and
then shifted on to somewhere else. One Sunday arternoon, when we come out
of Chapel, a big Steam Engine with 3 big vans stood on the side of the road.
It had broke down and had to be pulled into Mr Desborough's grass field by
one of his Engines. It was a travelling Cinematograph Show and it stood there
several weeks afore it was mended. When it was mended it gave shows for one
week. It had the Stage outside where the Clown and a Girl danced and rang
the Bells and with the organ playing it really put Magdalen on the Map.
Being Rabbit and Rat catchers we allest had plenty of Steel Traps (Gin
Traps) which we used nearly every day but which are now illegal to use on
account of being cruel, to which I agree. When I was about 7 or 8 years old
I would set Gin Traps for Rabbits at the bottom of our orchard, then about
7 o'clock I would git my little Candle Lantern and go round to see if I had

any luck. If I had got a rabbit in one I had to go and fetch Father to kill it as he could give it one twist and break its neck whereas I had not the strength in my wrists to do it so quick. Popguns med from Elder Wood, Whips and Tops, Marbles, Hoops and Catapults are nearly a thing of the past but they was all things that we med ourselves and played with, when we went to school. Whistles could be med with a bit of Elder stick, as well as arrows med from a strong piece of dried Reed tipped with a short piece of Elder to shoot from our bows. A Peashooter could be med with Elder and a bit of thin steel spring from an old pair of Mothers corsets. Bird Traps med by using four Bricks and 2 crotched bits of willow and baited with Bread Crumbs was another plaything, or a Cinder Riddle, and a piece of stick was another way we had to catch birds. You might say "Fancy catching poor little birds", but we was glad to have them in Sparrow Pies. As I was saying afore about that Old Eager Tide that used to come up the River, it was a Wall of Water, 5 or 6 foot high, and would come as fast as a Hoss could gallop. Old men used to talk about it when I was a little old Boor and would reckon Dates up by sayen "The Time of the Eager Tide". Then when I was about 10 or 12 years old, Blast if it dint start acomen agin and I believe it do now. Its a rare sight as it roars along and waves nearly turn the little old boats over. Another thing that was in Magdalen when I was young was a bit of ground fenced off and was called "The Pound" and was where any Hosses that was found loose on the roads, was put, and the owners had to pay about a Tanner to have the gate unlocked. Father used to say there used to be another bit of woodwork called the "Stocks" but it was gone afore my time. He has often said anybody who was helpless drunk or had bin nocking his old woman about, was put in the Stocks for perhaps half a day and set with his hands and feet through holes med-a-purpose and pad-locked in. People was allowed to chuck eggs or Tater-peelings at the one who was locked in. It was right aginst the Church Wall so anybody who went to Church could see anybody if they was locked in. Mother used to laugh about Old Mrs Dardy Place who hes took some Sunday dinner down when her Old Man was locked in the Stocks and fed Dardy with a Spoon.

I hev told you all about our house but I mustn't forgit to tell you about our other little house. It used to be called by a lot of different names some on'em I don't mention. *It was the Privey* or *Petty* or as my old Godfather usta say when he tried to talk posh *"The Double O C"*. It was a joint affair, one side belonging to our neighbours and the other side to us, back to back. Although you couldn't see into theirs, you could hear all wot was going on and it was a regular thing when Father and Arthur Bugg next door was both in their separate *rooms* to talk about the weather or any other bits of news they had heard. Inside was a Big Seat for grown-ups and a small seat for children. The door didn't fit very well at the bottom so it usta be a bit Drafty, the Privey standing about 50 yards down the garden path. In summer it was covered over by a creeper plant and shaded by a Bully Tree but in winter, when the leaves had done, it looked (and was) a very bleak affair. To empty it took several days of arranging, both cesspools would be emptied (allest at night) into a big hole and covered over with long straw. It would be left for a few weeks then Father would buy a few bags of new Lime, and put in it as well as plenty of chicken Muck, and perhaps a bag of Soot, and in time the new Lime would dry all the contents up. Arter it had bin turned over several times it would be ready to sow on Strawberry plants, Raspberry Canes and Currant Bushes and as Father used to say "Yow can't beat human Muck".

As we lived nearly opposite the Church we allest had plenty of Music – The Bells ringing, Organ playing, and Choir singing, in fact sometimes a bit too much, as at New Years Eve the Big Bells would ring the Old Year out and the New Year in, and would ring then every hour until 6.00 am. I often think its a wonder we dint catch all sorts of deseases, what with the various Stinks, sometimes River Water to drink or at the best of times, rain water off the tiled roofs which arter a long dry spell would be moss-grown and covered with Filth, but we was allest healthy, all Breast fed and as tough as nails. We had nearly a mile to walk to school, no hot school dinners in them days so home to dinner and back agin to school untill a quarter to four. On our way to school we passed the Blacksmiths Shop where there was allest Hosses inside and praps 8 or 9 outside waiting to be shod. Past the school towards the Fen was the Water Engine, Steam driven and it had a huge Paddle Wheel. This drained the Fens, the water being forced into the River Ouse at a big Sluice, quite near the School. As the Church Yard was very rearly used for funerals, a Cemetary was opened in 1900 and Father was caretaker and grave digger. If it was high tide in the river and the Water Engine was pumping graves would be half full of water so that meant trying to keep the water in the grave low untill the Corfin was ready to be lowered then a few handfulls of saw-dust was chucked into the grave to make it look dryer.

FEN DIALECTS

The old time Fenman hed a language of his own, lots of words he used was not heard on only in the Fens or praps up Norfolk. Very often now I let a word or two slip out and quite likely the person who I am talking to don't know what I am on about. Only the other mornen I was visiting some friends who was Lincolnshire people and I hapt to say "There must hev bin a lot a rain cos there is some big Swidges on the road." They both looked at me and din't know what I was on about until I said "Some big Puddles". I do think one word stand out afore all others and that is *A Rumman*. That word is used with scores of meanings. If it was a rough day people say "Thas a Rumman ent it" or if someone git hurt or someone died its still a "Rumman". Anything which they could not understand or exciting News was "It Sartenly is a Rumman". *Boors* and *Mawthers* (Boys and Girls) and Wenches for grown up girls was allest used and any body who was tall and thin was allest *Lanky*. Boor was a rare word used to welcome anyone even strangers. It would be "Coo Boor, you han't altered a sight" or it might be if it was a Stranger "No Boor, I dont know yar, you must be a furriner to me" perhaps coming from the next village. If in a crowd someone's toes might be trod on then it would be "Cor Blast, you hev now jamed on my Toes". A Horse was allest a *Hoss* and a Donkey was a *Dicky*. A Dog was a *Dorg*, a Heron was a *Harnser*, A Shrew was a *Ranney*, a Vole was a *Bog*. Rat and Snails was *Hodmedods*. Blackbirds was *Blackies*, Hedge-sparrows was *Hedge-Bettys* and Thrushes was *Mavises*. If anyone was a bit foolish they was *Duzzy*, and anybody who was active and small was a *Botty* little person. To hev a walk round praps down an old Fen Drove or Loke, was heving a *Ship Winder Randy* or it might be "Heving a Mooch Round". Childen rawing (crying) was told "To Stop that Blaaing" or if they was cold they would be told to "Hap yourself Up". *Hap up* is used every day now on Fen Farms etc. as at the end of the day it is allest "We gotta Hap up yit?" meaning make everything safe for the night. A boil on the neck was allest a *Push* and a Sty on the eye was a *Stainer*. Anyone who was talking (running on) silly had a *Lot of old Slaver*. If a person wanted to make anyone hear who was some distance away they would *Holler* and to shiver with cold was to *Dudder*. A few days ago a friend of mine called and he had a friend of his with him in the car. I told him if he wanted any Raspberrys they could pick some. I left them picking so they come to me with about half a chip full. I asked him "Haya got anew?" and him being a Fenman said "Yis. Thank Yow" but his friend din't know what I was on about untill my friend explained to him it meant "Have you got enough". Years ago if someone said they had had *Swimmers* for dinner it could mean they had "Light Dumplings" and I might add there was a big art in making them. My Mother had a big iron *Biler* that usta hang by the handle on to a Hake that was built into the chimney so that the Biler hung over the fire. Inside the Biler might be a Rabbit or half a Bullocks Head and carrots and onions, then at the proper time she would drop 10 Dumplins (often called "Twenty Minute Swimmers") and I can tell you there wont a better dinner anywhere on a cold day. Anybody talken quick would be *Chittering* and a lot of old rubbish laying about was "Old Kelter" or if someone was rushen about they was "Clawing about". If a little childs nose was dirty the Mother would ask "Hen't you got a Hancher" (handkerchief) and anything not cut straight was "Cut on the Sosh" and paths was *Footpads*. Looking at a Picture Book was looken at the *Gays*. Anything soft was *Fuzzy* or if was brittle, it was *Spolt*. Persons who was short tempered was *Titchy* or *Ungain* or people might say "They had got Old Harry (The Deval) on their back". A Servant girl was a *Skivvy* and a bloke who done odd jobs on a farm was "*A Sort of Laky*". There was one thing a proper Fenman would allest hev in his pocket and that was a *Shet-knife*. They was used for you might say "A Hundred and one jobs" such as Gitting thornes or thistles out of the fingers, Stones out of hosses hooves, cutten a lump of Bread and meat or cheese at Docky Time, trimming finger nails, cutting Broches (Stack Pegs) for thatching and in Fathers case to cut Moll-sticks or Pegs and skinning Rabbits and Molls. Perhaps a pig won't be dewing to well, then out would come his Shet-knife and Father would "Slit its ear" or cut a bit off the end of its tail to *Bleed it*. All the knives Father used he bought from a Pawn-brokers shop in Kings Lynn, which had been pawned by Seamen and had never been redeemed. They was allest kept as sharp as razors so he used to say "He could shave with them". In them days if a gang of men stood in a bunch when they should hev bin working it was a rare saying "I bet they are swopping Shet-knives". Before I finish about Shet-knives I must tell you a short tale about my early days. As I hev mentioned before, Father and my Godfather was Rabbit and Rat catchers (Father was also a Moll Catcher) and I was nearly allest with them. Perhaps we had finished ferretting one lot of holes and was ready to move to another lot. Before they moved on they would *Hulk* (Slit down the Belly and pull the Guts out) the rabbits which when there were several rabbits would make them much lighter to carry. When they had finished "Hulking" my old Godfather Bluff would say "Time we had a bit of Tommy (lunch) eant it Ted?" So out would come Bluffs Bread and Cheese, he would slice it across the middle and say "Here are Boor, git this down yar" and his hands and knife was jest as he had gutted the rabbits – they hed not bin washed. But I don't think I should enjoy it so much now as I did then, but we was not brought up squeamish. Still on about Shet-knives, several years ago a very old man died at Coldham (near Wisbech) so someone measured the old boy with a bit of string (Length ways and Cross ways) and took it to the Undertakers at Wisbech for him to make a Corfin. When it was ready the Undertaker and his lad assistant put it on a handcart and pushed it to Coldham (about 5 miles). When they went to put the corpse into the Corfin they found this old Fellar had died with his knees up so they could not git the lid on. Out came the old Undertakers knife, and he soon cut the leaders under the old chaps knees and so they straightened his legs and everything fitted lovely. On their way home they stopped at the "Chequers" Public House at Friday Bridge for a drink. The Undertaker pulled a lump of Bread and Cheese out of his pocket, sliced it across with his knife and give the Lad half and eat the other half hiself, and the Knife had not bin Washed so now you can understand why I eant very keen on Bread and Cheese. If anybody was Tall and Slim and walked with a slight stoop he was a *Slamakin sort of Cove*, or if anyone dragged their feet along they was *Slouchen* about. The smallest pig in a litter was a *Racklin* and a stack of Straw or a pile of Boxes that was leaning was "*Boughin Out*". A Person with tender feet walking along the road was *Pampling about* and anyone shouting would be *Yorping*. A piece taken out of the roadside to drain water off was a *Grup* and to hev a Tickling Corf was *Tisicking*. Fat persons was *Jot Gutted*, a poor eater whether it was Man or Beast was a *Pingler*, to be sick was to be *Gaggy*. A Persons big Toe was allest the *Tom Toe* and someone wearing a Finger Cote was wearing a *Hudkin*. Although there are many more old Fen names these are jest a few which are rarely heard today. If someone said "I hean't hurd a Dean" it meant they hadnt heard anything, and anyone suffering with Influenza had got "The old Hin flew outa tha Windar". Rheumatism was allest the *Screws* and ear-ache was tha *Lug Ache* and anything messy or sticky was a *Slarry old Mess*. To *Chuck* or *Hull* a stone meant to throw a stone and anything not very fresh was *Fusty*. Roads that was Frosty and Slippery was *Glibby* and four was allest pronounced as *Fower* and Two as *Tew*. To exchange was to *Swop*. Anyone with *Plenty of Yap* was a person who had plenty to say for themselves, sometimes it was *Plenty of old Rattle*.

BIG CHANGES

I very often think as I walk about the Fens, doing my job of Mollcatching etc. what big changes has come about since I was a little old Boor. Now a days its all 'Tinned Music" where years ago it was "Straight from the Singer". Very often I meet somebody or praps a gang worken, and there beside them, blaring out, is a small wireless set. I hev seen someone walking or riden a bike, with a wireless turned on giving the latest Pop Music or Cricket scores. The songs now a days havn't got any meaning to them, not like the old Songs of 60 or 70 years ago. Songs like *"Old Jims Xmas Hymn"*, *"A Bird in a Gilded Cage"*, *"The Volunteer Organist"*, *"Don't go down the Mine Daddy"* and scores of old songs that had somethin about them that would bring a lump up in the throat, and the singers never had a *Mike* to sing into to make people hear. There isn't the Art now in Singen as there was years ago, all they do now is stand there twisting, swayen and shreaking about trying to imitate American slang so when they hev finished you don't know what they hev been on about or if they hev had a bad attack of Belly ache. There didn't used to be any "Bingo", Whist-drives was considered "Dens of Vice" in fact I was not allowed to go to a Whist Drive until I was nearly twenty. Father used to let me go to a Dance as often as I liked but he used to say *"Them Twitch Drives ain't nowt only Gambling Dens"*. When we had our Harvest Supper there would be a Dance with music played by Loggens on his Accordian, accompanied by Billy Sizer on his Fiddle, and there would be the Barn Dance, Two Step, Lancers, with a Horn-Pipe sometimes for a change. Then someone, mostly Old Billy Dobson, would get up and give a nice song, then the glasses would be filled up agin. Everybody enjoyed theirselves but nowadays it won't be classed as anything at all. Even fruit growing isn't the same, what used to be wonderful Apple and Pear trees are never heard of now. About 1909 Father bought an Orchard down Magdalen Fen and it was full of big old trees such as Prune Damsons, Medlars, Quince, Bullaces and Russet and Beefam Apples, while to gather some of the Pears took 2 long ladders lashed together because they was so high, but they was lovely big Pears when they was fit and I hev never tasted Pears like them, they was like honey. As the old trees was so tall Father allest med a bargain when he sold the Fruit that the buyers had to come and gather them two Big Pear Trees theirselves. Which they did. He allest sold to a very old Lady (Mrs Exley of Lynn) and her two Nephews, Jack and Ben Culy, would come over and pull over a ton of Pears off each tree. Another sort of Pears we growed was a very early sort, Father would pull them when they was hard and green, put them in barrels which had green stinging nettles top and bottom, let them stop there about a week and then when they was took out they was a lovely golden colour and quite soft and they was ready for sale at Kings Lynn Gala (about August 3rd). At that time there was only about 2 sorts of Strawberrys, Paxton and Royal Sovereigns, but now Paxtons which was a lovely flavoured Strawberry and very firm is never heard of and there are not many Royals grown. People nowadays dont give the attention to Strawberrys that they used to, everything is done for quickness. Then plants was kept separate, so it could be hoed right round it, runners that form young plants would be cut off and when being strawed all straw had to be put under all the fruit vines so that the fruit rested on straw but now its just sprinkled on with a fork so half the fruit is on the earth, then they wonder why they get so many rotten or slug eaten strawberrys and if its a wet season the fruit are splashed with mud. We used to pull them for a farthing a pound but in those days anyone could buy a 6 lb chip full for a bob but its like everything else today its all done for quickness they call it *progress*.

Every year big gangs of all types of people would gather in this district for the Pea-Picking season. Word would get round that some of the big Farms had Peas ready so tramps, gypsies, and of course local people, would be there at a given date to make a start. The price for pullen was 1/- (old money) for pullen 40 lb of Peas but plenty of local people would earn £2.0 a day (5.0 am–7.0 pm). Some of the Tramps would sleep under lumps of Pea Straw and I well remember being in the Pea field at 5.0 am one morning and had jest med a start when a lump of Pea Straw started moving and a man crawled out. He asked me the time and told me he had half a bag of Peas left from the night afore so he slept out with the Peas under a lump of straw. Now there isn't a lot of Pea-pullen done, its all done by huge viners that pull, shell and sort them out and in a very few hours they are in tins. Taters now instead of being picked by hand are Picked, Bagged and Graded by Machines, and you might say more than half of them are put in tins. Machines hev even been invented to pull Strawberrys but at the moment they don't seem to be getting on too well but no doubt the day will come when they do it. It crossed my mind the other day anyone could go into scores of homes and not see a Spill (pipe lighter) on the mantlepiece. When we was kids we had to twist paper to make Spills or sometimes we would get a nice bit of wood which could be split easy with a knife then we would make a hul lot and they would be stuck in an old Toby Jug on the Mantlepiece so all Father had to do was reach up, get a Spill, hold it in the fire for a moment and light his Pipe. Wooden Spills would last for 3 or 4 lightens where as paper Spills won't only last once. Its years now since I last saw a Grill that used to hang on the Fire-bars of a stove. Bloaters would grill lovely with the grease sizzling out and droppen into a tin Pan underneath them. The smell of Bloaters was lovely and it make me think of a yarn a funny old bloke told me of how he got his old woman back. Arter they had had a row she left him and the Kids and cleared off. She was gone a couple of days but on the Saturday the old boy had bought some Bloaters and was grilling them for his and the kids Teas. One of them (a girl) looked out of the winder and said "Dad theres Mother outside". So he say "Let her keep there". Then she knocked on the door so he say "I went and opened the door and she say 'Cen I come in, I am hungry". He say "she come in and Blast if she dint eat two hul Bloaters, so that want me she come back for, it was the smell of them Bloaters that brought her back". This same old bloke would allest finish his sentences off by saying *"Blast Boor eint it a Rumon"*. My Father would allest finish off what ever he was saying with *"As Paddy Say"*, Old Bluff with *"Bless my Heart"*, Bill Lee with *"Withererno"*, Bill

Bates who kept the Galloping Diky Pub with *"Tickle my Britches"*, Old Addley Kabble with *"Lawk a Mussey"* and Old Charley Benefer, who had been an old Sailor, with *"Yow talk like a froggy"*. Since the two Wars, and most of the oldens gone, its getting a job to find a true Fenman, corse people hev come from all over and even foreigners hev settled down in the Fens so if they ask what the time is and you say *"Half arter fower"* they wonder what you are on about. Yes there are big changes and there isn't the same type of persons about that there used to be when I was a boor 70 year ago. One oldish woman stand out in my mind, she used to come from Kings Lynn with a big gang of women to pick Apples and Taters at Harrison Bros big Fruit Farms at Magdalen. She was a big, tough strong woman and her language would make a London Docker scrivell up and she didn't care about anybody, but I hev often been told *"She had a heart of Gold"*. We had a Baptist Parson (Rev. Jackson) who used to preach at Magdalen once a month and the Rev. used to play Football for Lynn Town. Also playing for Lynn at that time was a player called "Mary" Earle and was the Son of the lady I hev jest mentioned (*Annie Earle*). Both Mother and Son lived at what was called the "North End of Lynn" and the Rev. Jackson knew them very well as Mary played football with the Rev. He has often told us that although they was rough tough people they both had "Hearts of Gold" and if there was any trouble in the Football Field and perhaps Mary was going to fight somebody, he had only to go and get hold of Marys arm and he would calm down and he said "If I wanted any help of any sort both Mother and Son would be the first to offer." A friend of mine (Miss Kate Woodhouse) lived in the same street as the Earles when her mother died leaving a large family and she has often told me that Annie was wonderful to them and allest made it her business to know if the children wanted anything as regards food, clothes, etc., and was like a second mother to them although she had a big family of her own. Another old character was old Stuttle Fox. He had been a Ships Carpenter for over 40 years and when he left the Navy he had a Carpenters shop at Magdalen, right close to the "Cock Public House". He was a wonderful tradesman and could make Carts, Barrows, Coffins, in fact Stuttle could make anything but he was not to be depended on, because when he was paid for something he had med, he won't do a lot more till he had drunk it away. One eyed *Billy Hart*, who never had a home only somewhere where he could jest draw in for the night was another character. He would travel with Mart and Fair People from one town to another selling "Song Sheets". He allest wore a Hard Hat and for a penny would sing a Song which went more like this *" Yow be my Wife an I'll buy the Ring, With Sarvants to wait on yer, yow pritty little thing"*.

Walking in Wisbech last week I saw a Big Motor Dray unloading Beer outside a Pub and it flashed acrost my mind what a change there was since I was a little Boor. I remember watching the old Hoss drawn Dray stacked up with big wooden barrels, the Drayman wearing a water-proof apron, busy sliding barrels down to the ground then rolling them towards the Cellar. Only a short time ago a friend was telling me about his Grandfather who had been a Drayman all his worken life. He said Grandfather used to leave Wisbech every Monday Morning and would not get home agin untill Friday Night. He had a double Shafft Dray but when the roads was bad in the Winter time he had three Hosses, one in front of the other two. He said arter he left Wisbech he was never Sober until the next Saturday, because at each Pub he delivered Beer there was allest Free Beer and $\frac{1}{2}$ oz tobacco. He arrived at Ely late on Tuesday night, stopped there all Day Wednesday, then on Thursday Morning he started for home, delivering on the way by a different route. One bit he told me was one Friday Night Grandfather never arrived home but when Grandma looked out Saturday morning it had bin snowing and there was a heap of snow jest inside the Garden Gate. She went to see what it was and jest aforo she got to it, it moved and then she saw it was her old man. On the Friday Night he had bin drunk as usual and that was as far as he got, but the funny thing was he never missed a Pub and allest managed to put the hosses up and feed them before he went home. My friend said if his Grandfather went to sleep on the dray his Hosses knew jest where to pull up and when he had three, the front Hoss would go to Ely and back without even heving a rein on him and if the Hoss saw another cart coming towards him he would pull to the side of the road to let it pass and perhaps Grandfather was fast asleep. Fifty to Sixty years ago Oil (Paraffin) used to be taken all over by Hosses and an old Boy named Erne Humphrey of Lynn has told me that he used to drive a Tanker and two Hosses and deliver oil round Norfolk as far as Dereham. Pop was also delivered by Hosses and lorry but the Beer Drays and Oil tankers used heavy hosses (Shires) the Pop Drays used what we called half Strained ones (trotting hosses) but there was one with 1 Hoss and one Mule, the Hoss only working every other day gut the Mule every day because the old Mule could stand it, while the Hoss couldn't, so the poor old Mule done the work of two Hosses.

Nowadays if there is a Fire people ring up the Police or the Fire Station and in a very few minutes huge Fire Engines are at the Fire, pumping tons of water, quite different to when I was a boor. One particular Fire allest stands out in my mind. It was about 4.0 pm and some people was leaving off work, others still at work, when a loud banging was heard comen from the direction of the Fen. At last we saw the cause of the banging, it was Old Music Lake in his Donkey and Cart and he was a banging away on a big Tin and shouting out *"Fire, Fire"* and pointing over his shoulder in the direction he had come from. The old Donkey was going at *Top Speed* (about 4 miles an Hour) and Music was on his way to Watlington Post Office to phone for the Fire Brigade from Kings Lynn. It was at least another hour afore the Hoss-drawn Fire Engine arrived, but everyone done their best. A lot of people who had heard Music went to help and all live Stock was saved but the Buildings and Stock Yard was burnt up. Poor old Music, who was very deaf and used to shout very loud, often said arterwards "Old Moses (his Donkey) seemed to know it was urgent as he had never galluped so fast in his life". Then his voice would drop and he would nearly whisper "He won't never hev stood it if he harn't a bin a whole-un" meaning the Old Donkey was a Stallion or Jack-Donkey and hadn't been castrated.

FENLAND BYGONES

Afore the two Wars people never hardly ever travelled far, in fact people living in the next Village was classed as foreigners. In my time there was Men and Women living down at the bottom of some Fen Farms who would only visit our Village about twice a year. They doctored themselves and their Cattle, med their own Bread and Beer, med their own Shirts and underwear, and very often the Men still wore the same Sunday or High Day suit as they was married in, haps over 50 years afore. My Mother med Bread, Beer, Shirts, put linings in trousers (we never wore pants), darned the socks for our family of Father, Mother and 13 Children. Father only went to school one half day and he couldn't read nor write and my old God-father Bluff was jest the same, never han't any Schoolen. Wellington Boots han't never been heard of and Men and Women working on the land wore leather boots and wore bags (sacks) on their legs from Boot tops up to their knees. There was a Relief Officer that used to come round once a week from Downham and give Widars and old people who was too old to work a half a crown and a Bob each for any children, cos there was no Old Age Pension in them days. There was no tap water, so rain water was caught in tanks and tubs and if there had been a long spell of Dry weather water had to be fetched from the River Ouse to drink and wash with. When the dykes got low a big hole had to be dug in the bottom of a dyke so the water could drain into it and the cattle's drinking tanks would be moved to the side of the hole and the tanks filled up by using a Jet, which was a sort of bowl on a long handle. Very orfen the water was as black as soot but the cattle had to drink it or go without. Mens wages was 14 or 15 Bob a week, the days work started at 7.0 am and finishing at 5.0 pm, for six days a week, no Saturday afternoons off then, no holidays and only Christmas Day paid for not working. Good Friday we used to finish work at

12.00, but on some farms men used to hev to work till 10.00 am, go to Church, home to dinner, then back to work at 1.00 pm. When I started work on the land in 1913 I got 9 *Bob* a week, I was a strong Boor. I did just the same work as the men but arter I started working with hosses I got a bob rise (10/-). There isn't the hard work on farms now like there used to be cos they got Machines to do all heavy jobs like Muck Fillen, Muck Spreading, Dykeing, Digging, Pitching Shoves of Corn, Corn carrying (18 stone), Beans (19 stone), and artifilar Manure (16 stone). All these weights had to be carried on Mens Backs and very often up Granary Steps. Hoss-men had to be in the stable at 5.0 am to feed the hosses and they was not allowed to leave the stables before 7.0 pm, arter they had racked up for the night, and that was arter walking all day on hard clods on a bit of Fallow land. Now-a-days Tractors are used instead of hosses so even if it is a bit bumpy the men are riding and if it come on to rain they hev a shelter over their heads and when its time to leave off work, they ride their Tractor home, switch it off and away they go home and no Racking up (Feeding and Bedding down) to do. I never thought I should see sich changes in the countryside as wot I hev, and it make me sometimes wonder how we used to manage afore Electric, Gas, Petrol, Radio and TV ruled the world. When I was paid my wages of 10 Bob a week, I did get somthin that you dont see now very often, and that was I was paid in gold as paper money had not been heard of then, so I used to get a *Gold half Sovereign* for a weeks work.

CHANGING SCENES

Very often when I think of things that hev changed during my Life time (75 years) I can hardly believe so much could rearly happen or that such things did usta happen. How well I remember the Elections of 60 to 70 year ago when it was either Conservative or Liberal round our way, who stood for the Election. In them days there would be amusing Posters hung up by each Candidate, one or two that I still remember. It showed a Liberal (Lloyd George) shooting at an old Pidgeon setting on a Tree and said

"Lloyd George bent his Bow
Aimed at Pidgeon and hit a Crow"

and another one was

"Lloyd George took his Bow,
A Russian Bear to find,
But when he saw it, ran away,
With *Speed* and *Bare* behind" (The Bear had torn
his Trousers backside)

and a third one I remember was,

"What is the good of a Small-Holding to me,
Without Tarriff Reform"

and showed a ragged old man ploughing with a Donkey.

I remember the colours was, *Pink* and *Purple* for Conservatives, and Blue and White for Liberals.

Of course there won't any Television or Wireless in them days and it

Fen dyke siding tool.

might be a week afore some of the people living down the Fens knew the results of the Election unless of course the Conservatives won, then the Boss would tell them, as nearly all the Farmers was Torys and on some farms the Farmers would tell their labourers how to vote, cus a lot of the old workers couldnt read or write. Our Postman allest biked except for a few days afore Xmas, then he had an old Hoss and Cart to bring the letters etc. for Saddlebow, St Germans, St Peters and Magdalen. He would arrive at Magdalen Post Office about 8.0 am, hev his breakfast in his little hut, then deliver letters all round the outlying part of the Fen and when that was finished he was done untill about 4.0 pm when he would collect from two or three post Boxes, pick up at the Post Office and leave at 5.0 pm and pick up Mail etc. all along the route to Lynn. During the day he would do Boot and Shoe repairing in a Small Hut next to the Baptist Chapel. At that time we lived close to the Post Office and hev seen Freddy Amens (the Postman) set off when its bin Snowing fit to Blind yar but all Freddy would say is "I shall hev my Tea, baked, boiled and fryed by the time I git home to night". Another thing we never see nowadays is women walking along the hedges or the River Banks hunting for Sticks or Firewood. They would hev their old Corse Apron (Made with a hessian sack) full of sticks to light the fire or if they was lucky on the river banks they might find some decent bits of wood that had floated up with the tide. Then when the Tides was right a gang of them would go up to Stuttle House Corner where the old River Ouse usta run in a big curve and there was sand over half way across the River. It was allest called the Coaling Shore because how ever much was dug up with a fork or rake there was allest plenty more. While the tide was running down the sands was hard to walk on – but as soon as the Tide turned and started to run up then was the time you had gotter git off or you begun to sink in. Old Mrs Bugy Leverington had an old boat and she would take the other women and their bags of coal up to the steps aginst Magdalen Bridge and she would charge them about 2d or 3d (old money) each. Our Policeman also biked round (or walked) except when he had someone to take to be locked up at Terrington Lockup, then he would git my Father to drive him and his prisoners to Terrington, in fathers Hoss and Cart. The Supt. of all Marshland Police would come round in a Hoss and Trap and he allest brought a young policeman with him to look after the Hoss while he talked to the Village Policeman. The fust Policeman I cen remember at Magdalen was a huge man, his wife was a very tall woman and his children was all taller than other kids their age. His name was Woodcock (we used to call him *Timber Diddle*) but of course din't let him hear us. I hev seen him biking down the old Fen droves, the wheels of his bike one mass of Mud but he was allest happy. He was more than just a Policeman, he was a friend to all and the kids loved him. He liked his Pint (two better than one) and there wont anything he liked better than to set in the "Lodes Hed" or the "Dolphin" Pubs and get arguing about the "Bible". He knew the Bible from start to finish.

Another thing of the Past is the Steam *Thrashing* Machines moving from one Farm to another and as we lived then right close to "Happy" Desborough, who owned several Thrashing Sets, it was a regular thing to hear him moveing his machine late at night so as to be ready at another farm about 6.0 am next morning. I usta think it was a lovely sound on a nice still day to hear the thrashing Drum humming away, perhaps every now and agin hearing her give a bit louder groan as the Band Cutter had let one go through without the Band being cut. The smell of the greasy smoke was allest something I used to enjoy and if by any chance I smell something like it nowadays my mind go back to the old *Threshing Machines*. Then there was the Steam Railway

Engines, which when starting with a heavy load would send a long column of smoke up into the sky and blow steam in clouds before it got its heavy train moving. There was something about those old Engines that is missing from the Diesel Engines and although the Driver and Firemen would sometimes be as "Black as Tinkers" they still loved their old Engines and treated them as if they was Human. Harvest Men and Women tying Corn in the Harvest Fields from early morning untill late at night and then very often having to walk to and from work was tired but happy. It was very very hard work for very small Pay and tying corn properly was a Work of Art and I should know cus I learned to walk alone for the fust time in a Harvest Field. As soon as us kids was old enough we had to make *Bands*, first of all for Mother, then as others grew up, for Father. Bands was med with two handfulls of Wheat, Barley or Oats. Just as it was cut which had to be twisted a certain way so they held together then laid at the bottom (arse end) of the sheaf and the tyer would place the sheaf on the band, get hold of each end of the band, pull it tight, give the two ends a double twist and tuck the ends under the band. Later in the day the sheaves would all be picked up, one under each arm, and stood up with the grain end uppards with 10 or 12 sheaves to make a Shock. There they would stand untill they started Carting. The Shocks would then be pitched on to wagons and taken to a big stack and if they was not to be threshed they would be Thatched to keep them dry untill perhaps the Spring. After the fields was cleared then along would come the Gleaners, which meant picking up the stray ears of corn, and putting them into sacks, which after being threshed by hand would go to the Millers, and be ground into flour and so save the House-wife having to buy Flour from the Bakers. After grass had been cut and made into Hay it would be carted Loose and med into a stack where it would heat slightly and so get a nice nose or smell unless it was to damp when it would go "Musty". After the stack hed settled and the top put into shape, it would be Thatched. When it was wanted for the Cattle during winter a cutting would be med by using a Cutting Knife (a big blade at right angles to the Handle) and a man who was used to cutting would cut a nice square Fleck ready to carry on a Two Tined Fork to the Cattle Yards. Now its cut, baled up by machine and very seldom has any nose to it but it does save a lot of hard work. In our house, Father being no Scholar and Mother not very educated, we din't hev a daily Paper, just a Sunday Paper, and the "Lynn News" on Fridays and the Police News on Thursdays. Us children had to read the papers to Father and Mother and the other children would be making "Pegged Rugs". The "Police News", was full of Crime of every sort but more especially Murders. It sometimes showed the Murdered Persons with their Heads cut off or their bodies ripped open, awful pictures rearly, and the Identity Kit we now git on Television is miles behind because the Police News always showed the Murderer doing the Murder although no one had ever seen him. If there was a execution, there would be a drawing of the Murderer standing with the rope round his neck and it got so awful that Mother used to dream or have Night-mares about them so at last Father gave up taking the *Police News*.

Besides the Coal Man coming round we used to hev the Turf Man bringing blocks of turf (Peat) from Bardolph Fen. It was cheaper than coal, but did not give out the heat that coal does and made a lot of dust and dirt but it being slow burning if the Fire was banked down with turf there would be a fire next morning after it had bin poked about a bit and given a few puffs with the Bellows. When Mother was goen to start maken her dough ready for Baken one of us had to run up to the "Three Hoss Shews" pub to see if the Brewers dray had bin, if it had we would scuttle off hom agin and fetch a big basin which they filled to the top with Brewers yeast for about 1d (old money) so that was a bit cheaper than the German yeast which was sold by the ounce. When Mother did get Brewers Yeast we knew what we should hev for dinner corse Mother allest med enough dough to make us some dumplings. In my early days every Corn field had to be mown round before the Hosses and Reapers could start and nearly all Peas was cut with a Scythe. It was a rare sight to see perhaps 8 to 12 men mowing, one just in front of the next one and when they got to the end of the field it would be "Sharp Up" and every man would reach behind him where his Rub-stone was in a leather bag fixed to his belt. Now I should not think there is one young man in twenty who can sharp a scythe much more use it, but in them days they took pride in mowing.

All Artifical Manure had to be sowed by hand so men would carry a container called a "Sidlop" which hung in front of him full of Manure suspended by a leather belt round his shoulders. He could then sow the Manure with both hands and so was able to do two rows at once. It was heavy work with bad walking as the sower had to walk on the top of the potato ridges, but most of it was Piece-work so he might earn 2 bob more than his days pay. Now its all done by Manure Sowers pulled by a Tractor or now its even sown in Liquid Form before the Taters are planted. Bullocks that years ago used to be out in the grass fields untill October, now sometimes never get a chance to see grass at all unless its for a month or two while they are Suckling. In 1915 I was feeding a yard of 45 big Bullocks in one big yard for a local farmer who farmed at Magdalen St Peter and Tilney. First thing when I got there at 7.0 am was to carry all the food into the big yard and put it into various Cribs so that all the Bullocks could feed at once because there is nearly allest a Master Bullock, one that Bosses the others. Arter feeding there would be water to carry by pails and yokes and pour into a long tank, then bedding down with straw, all had to be done by 9.0 am because as soon as they hed done feeding they would lay in the straw and Chew the Cud, and as the Boss used to say "Thats when they are doing their-selves a bit of Good". Next would come the hardest job, cleaning and grinding mangolds and breaking Linseed and Cotten Cake. Mixing the food up was a layer of chaff a layer of ground Mangolds, a sprinkling of Meal and Cake, and so on until the heap would be 3 or 4 ft high. The mangold juice would run into the chaff and it would be nice and damp at feeding time. Of course it was a seven day a week job and went on all through the Winter until April when they would be fat and ready for Market. Now one man look arter 100 Bullocks there is

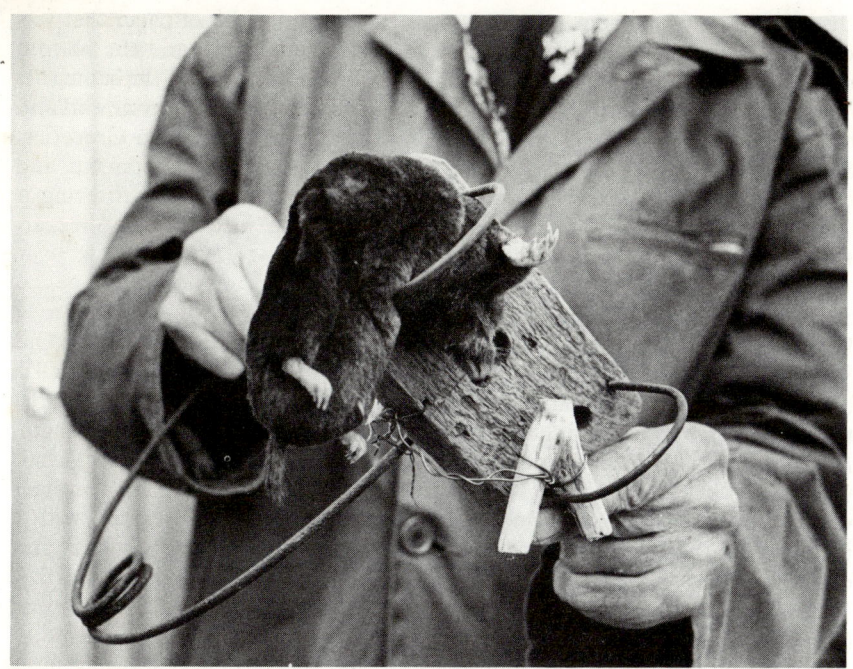

Wooden 'molltrap'.

no hand grinding, no Cake Breaking, the Straw and Hay is all in Bales and there are tanks of running water in all yards. A Machine go round all yards and fill big Hoppers or Cribs up with Crushed Barley, Cake Cubes or Nuts and Sugar Beet Pulp and Straw and Hay is carted into the yards by a Tractor and Trailor so all the hard work is done by Machinery. Its jest the same fattening Pigs, they are all fattened on Hard Fattening Nuts where years ago there would be a huge Copper holding a quarter of a ton of small (Chat) potatoes and this had to be boiled twice a day and three times on Saturdays. They would be mashed up, mixed with Barley Meal and water, then all carried in buckets into the troughs at feeding time. Jest Fancy 80 or 90 fat hogs rushing to you to gobble the grub up, you very often got knocked over or carried round the yard on their backs and the pails of grub sent flying. Now the pigman dont have to go into where the pigs are, but feed them from the outside and he has nice clean concrete to walk on. Even the job I now do since I retired from the Railway (Molecatching) is not the same, as my Father and Grandfather used to make all their traps, but now traps are bought at the Ironmongers and I dont suppose there is many people in the country who cen make a wooden Molltrap. The village Shop-keepers used to hev their Sugar come in Hessian Bags containing $1\frac{1}{4}$ cwt of Sugar, then they had to weigh it up into 2 lbs, 1 lbs and $\frac{1}{2}$ lbs and pack it into thick Blue Paper Bags. When children went after a penny, half-penny or even a Farthing worth of Sweets

25

they would be taken from a deep glass jar and put into a bit of paper that was cut into a square then twisted round the finger and thumb and the bottom twisted tight. Pepper and other small things would be put into the same sort of bags and weighed on Brass Scales with small Brass weights on one end and the articles on the other. The Village Shops sold everything from Groceries, Shirting Material, Parafeen, Toys, Cough Cures, Spades and Forks etc. and if they had not got what you wanted they would git the Carrier to bring it from Lynn the next day. Of course anyone wanting such stuff as Vinegar, Parafeen, etc. would hev to take their own bottles.

SUPERSTITIONS AND GHOSTS

People very often laugh now when I get running on about some of the Old Superstitions what their Fore-parents believed in, but I know very well a lot of them was true. If you had a visitor that got up and put his chair aginst the wall when he was leaving that was a sign he wouldn't be visiting you agin. Many is the time I hev heard Mother or Father say "No, leave that chair where it is" and Blacks (soot) hanging on the Fire-Bars meant "A Stranger was coming and somebody would claw up and knock the Blacks off with a poker. A Feather Bed med with "Wild Birds" feathers was very unlucky and old people allest said that anybody dying on sich a bed "Had a Hard Death". It was and still is considered unlucky by some people to bid aginst the Church (Parson). Two year ago I opened a Church Garden Fete at the village of Guyhirne near Wisbech in Cambridgeshire and the Vicar was showing me a lovely Silver Cross that he had bought at an Auction. I said to him "I bet that corst you a lot of money din't it?" so then he told me how he got it cheap. Two Dealers was bidding for the Cross then he put a bid in and the two men looked to see who it was bidden aginst them, then they never bid any more. Arter the Sale was over the two dealers come acrost to him and told him he had got a lovely bit of work very cheap, so he asked them why they never bid agin, and they replied "They never bid aginst the Church as it is very Unlucky". To put a Silver coin in a Baby's hand is to bring it luck through its hul life, but to cut a baby's nails is very unlucky and my Mother allest used to bite her Babys finger nails until it could "Runalone". A Hole in a loaf of New Bread meant a Corfin would soon be brought into the house and the same if a Robin come in or flew and hit the Winder. If egg-shells was chucked on the fire, the Hens would soon leave off laying and knives crossed on the table meant "Bad Luck". To burn Elder wood on the fire was also unlucky but if you had a Elder Tree growing aginst the house it prevented the house from getting struck by Lightening. Although 13 is allest considered to be an unlucky number, it is the best number of eggs to set under a hen. When I was a Boor, if people met a Piebald Hoss or a Cross-eyed woman and they spit on their Shoe it was good luck, or to see sparks at the back of the Fire was a good Sign and that you would soon be hearing "Some Good News". If you was outside when you first saw the New Moon and you turned your money over (Silver), that was a very good Sign *but* not if you saw it through glass. About the year 1909 a Man was drowned in the Great

Ouse near Magdalen Bridge. His cycle was found on the Mud Flats (between the river bank and the water) and his dawg that was allest with him layed near the bike but there was no sign of his body. Although men working on the banks near-bye kept their eyes up and the river was dragged by grappling-irons they couldn't find the body. Then a month or so arter two men, Shalley Egget and Jack Benefer (a local fisherman) got the old row boat out and took it in turns to row up and down the River all the time beating the Dig Drum that they borrowed from the Wiggenhall Prize Band. The next morning the body was found floating in the stretch of water where they had been "Beating the Drum" the night before. Its a very old Superstition that a body will rise to the top of water if a Drum has been Beaten above it I know thats perfectly true cos my Father served on the Inquest. If a person who has owned land where "Crows" has built their nests for years, die, then the Crows would leave their nests and it would be many year afore they started building there agin. This happened quite close to where I live. A big Farmer who owned a long row of very tall Elm Trees and although every year there was allest Rook Shooting the old Crows allest returned and built their nests and reared their young. After the Farmer died there was not a Crows Nest built there for nearly 20 years but now they build there in large quantity. Bees are thing that will swarm and leave their Hives unless told of a Birth or Death in the family and to swear at Bees will make them to clear off and leave their Hives. I hev heard Mother talk of the time Father come home the wuss for drink and Mother said "Them Bees are Swarming" so Father said "B---- the Bees" and every Bee left their Hives and never returned. Cabbage and other plant seed should allest be set on the rise of the Moon, if they are set when the Moon is on the Wain, more than half will be Bolters (Seedy ones). Shallots should be set on the Shortest Day and took up on the Longest. To get plenty of Double Bromptin Stocks, allest set the seed on a Good Friday. To see Sun Rays meant the Sun was drawing Water. Pigs should allest be killed (if for Bacon) on the rise of the Moon, if killed on the wain, the Bacon will go bad. (This belief is still obeyed here in the Fens.) If someone give you a knife or anything sharp, always give a small coin in exchange or the person will hev *Bad Luck*. I hev heard say that even our Queen allest insists on giving a coin in exchange for the Scissors that she uses to cut a ribbon when opening a New Bridge or a New Road, because a gift of Scissors can cut a line of Luck or friendship.

To spill Salt is bad luck unless a small quantity is thrown over the left shoulder, then its supposed to "Blind the Davels Eye". Walking under Ladders, Wearing Green, Peacocks Tails, and breaking Mirrors was all supposed to bring Bad Luck, but carrying a Lucky Stone (one with a Hole through it) a Rabbits Foot or a Four Leafed Clover are a few things that are Lucky. Very often when I was a little Boor I hev heard Father or Mother say when the Bell was tolling "Jest listen to that old Bell, its tolling for a Death" and I hev noticed very often we heard of somebody dying within a day or so. A gang of House Martins swooping round the Church, shreaking as they flew meant rough weather was on the way. To see or hear the Cuckoo before seeing a Swallow meant living all the year in Sorrow but its lucky to hear it on April 28th. There has been times when I was with Father Moll-Catching and I hev said "Look, there's a Foal" but if he han't seen it he would say "Which way is it facing" and he won't look if I said "Tail to us" but if I said "Facing on us" he would look and say "Thas Lucky". Not one Fen-woman in a hundred would do any washing on New Years Day cos that meant you either Washed one of the family out (Death) or wash one in (Birth). A room where a dead person is lying should hev any Mirrors covered and a small lamp or light left on. If a Pregnant Woman see a Hare coming towards her, the Baby might quite likely be born with a Hare-Lip. Red and White flowers should not be placed together in a room or Bad Luck is sure to follow. If there are 2 New Moons in the Month of May means it will be a good Hay Year, but to hear Donkeys Braying or Peacocks Shreaking means Rain and Rough weather is on the way. Its bad Luck to pass on the Stairs but good Luck if socks are put on inside out and worn like it (not to change them). If the Tea-Pot lid is left off it means a Stranger will be coming and to Sneeze and say "God Bless you" is Good Luck, but to sing afore breakfast mean to Cry afore Night.

GHOSTS AND THINGS

In the year 1900 a New Cemetary wus opened at Magdalen an my Father wus appointed Grave Digger, grass cutter, hedge trimmer, a sort of caretaker. He kept this job, which wus rearly only part time, cus in them days people wont allest a dying like they dew today, and also did his regular job of Vermin Distroying. It wus, I think in 1911, I know it wus the year we had Black Monday, when the rain poured down and never stopped for over 24 hours and destroyed thousands of acres of corn and drowned thousands of young pheasants, hares and other game. About a week afore the flood an old chap died and as Father wus pea cutten he decided that he would go with his mates cutten, then when he had had a rest he would go an dig the grave when it got cooler and he hoped tew git it dug about tew o'clock in the mornen and then go hom, go tew bed, an git up about dinner time and he would be ready tew bury the old chap at about 2.30 pm.

It wus lovely and moonlight but he took his old lantern and got busy. Somewhere about midnight he wus gitten on nicely but thought he would hev a snack so he got out of the grave and found he wont alone cus only tew graves away from where he wus digging wus a woman, sorta bending over a grave. The rummest part on it wus he recognised the old gal cus he only burred her a few weeks afore. Father said "I couldn't make this out" so he asked "What are you dewen on Addley" but she didnt speak or take any notice on him. He thought Blast this is a rum dew and he said he dint feel like eaten his bread and cheese, but still wish the old lady would say suffen but instead she started walken away and she went right through a thorn hedge, across a field under some walnut trees and on tew the river bank, an thas where he lost sight on her. Now old Addley wus a rum old gal when she wus alive, she could put the white mice (spell) on people, and the day she wus

Sparrow trap.

buried a stranger went down the cematary and arter the mourners hed gone he said tew Father (hew wus fillen the grave in) "You will hev a job tew keep her in" so Father said "Well Boor I heant lorst any from here yit". But after what he saw he wont tew shor. He looked at the place where she went threw the hidge but there wont nowt tew see and although the dew wus fallen he couldn't see where she hed walked in the grass, but he wus shore it wus old Addley.

Fifty one year ago there wus some strange things happened at Gorefield, near Wisbech, in a farm house, occupied by people named Scrimshore. Things went proper haywire, there was sich goen on there during the night, things knocked over, furniture moved and of all things an old Pianola would start played like mad but there wont nobody near it. Three men, one of them an old pal of mine (Laurie Howlet) asked if they could hev a night in the room where all this happened. They took some grub, plenty of baccy an settled down, but not for long cus all at once the old pianola started playen and it sounded as if a wagon and hosses wus goen round the house but there wont nowt there. Before my old pal layed down in a couple of chairs he took his change out of his pocket an put it on the mantelpiece but all at once it went flying all over the room. Things seemed to quieten down as it got daylight an Laurie and his pals decided tew hev a bit a breakfast. They hed took some sausages with them straight from the butchers shop so they got a fryen pan, pricked the sausages with a fork and started tew fry em, but thas when things rearly did go crackers, cus it dint matter what they done the sausages wont lay in the pan but stood upright.

When I wus a boy there used tew be a lot a barges pulled by hosses up and down the River Ouse. They would take loads up stream tew Ely and Cambridge and bring back loads for Lynn Docks. There wus allest a hoss on each bank, ridden by a young chap about 16–18 year old and the hosses pulled by over 100 yards of rope. The skipper or whatever title he went by, of one of the group of barges wus a big burly man, big enough tew knock a bullock down with his fist and as his name wus Jude allest wus called Judy. As a boy he lived on the river bank at Magdalen and he wus mates with my father an when they met they allest had a gossip. I have bin with Father molecatchen on the Ouse Banks up towards Denver Sluice and Judy hes give us a ride on the barges as far as Magdalen Bridge.

One day they got talken about if they wus late waiten for the tide and hed tew go down to Lynn in the dark. Judy said "We dont mind a bit cus the hosses know where they are and where the stiles and bridges are, but we ont pass the haunted house at St Peters (between Magdalen and St Germans) at midnight." Judy said, "The hosses will stop, and ont budge and sweat drip orf them," and he turned round and pointed tew a big, rough old dawg, what looked more like a bear and he sed "You want tew see him, he'll slink away an hide in the corner out of sight and make a whining row, and yit another time he would rip the guts out of a lion." Father asked him if he felt anything and Judy sed "Yes, but I can't describe it, ya legs seem like jelly". Fishermen who hev hed fishboats just near there allest sed at certain times figures hev bin seen comen from the haunted house and seem to come up from the river and walk in procession up the bank then vanish. Since then I hev thought a lot about that Haunted House an if a big tough fella like Judy wus scart then there musta bin suffin out of the ordinary cus Judy hed bin in the Navy and hed bin in some rough dews but allest managed tew fight his way out, but when he said "His legs wus like Jelly", I knew it wont nowt natural.

BLUFF AND THE PICKPOCKET

When I wus a small boor and right up tew a few year ago Kings Lynn Sports and Gala wus a very big event. It wus allest held on August Bank Holiday Monday and there wus Hoss Trotten, Whippet, Cycle and Flat racing and crowds come from miles around. The "Walks", (a park) cuss thats where it wus held wus an ideal place fŏrit and there wus amusements for the kids, an Beer Booths for the grown ups, so everybody wus happy. Among other things it wuss an ideal place for Pickpockets, an several of those light fingered folk used tew come down from London on the expresses and by mingling with the crowds of Country Folk done very well. For many years it wus the practice of the Police tew have a retired Detective tew be on Lynn Railway Station tew watch hew come off the trains. An old friend of mine, Detective Inspector Pedder, hew wus retired from the London Police, an on his retirement came tew live at Watlington, quite near Magdalen Road Station, in as he called it "Poppyland" (Norfolk), hed the job for quite a number of years as he knew so many of the Pickpockets and I hev heard him say many a time "He never forgot a face". He would stand quite near the ticket barrier and if he saw anyone hew he knew was "light fingered" he would give a signal tew the uniformed police, hew wus also near the barrier, and the gentleman would be detained at the Police Station until the evening an then put back on the train tew London.

Now my Father an my Godfather Bluff, med up their minds tew go tew the Gala an wus talken it over at their regular meeting place aginst Magdalen Bridge, an some of their pals advised them tew be careful with thur money as pickpockets would sure tew be there. As I hev said before neither onem wus any scholar an hed never bin far from Magdalen in their lives so the thoughts of pickpockets sort of scared em. Bluff asked which pocket would be best tew put their money in an the answer he got wus "They could git it outer yar mouth without your knowin about it". Now I jest said they wus rather scared but that won't quite right cus Bluff wont scared of anybody. He wus a huge man, well over six foot and weighed over sixteen stone and in a rough house could more than hold his own and it wus often said "He could knock a bullock down with his fist". That night Bluff come down tew ours and him an Father knocked their hids together and thought of a plan how they could go tew the Gala an still be able tew keep their money. As I hev said before Father an Bluff kept anything between 20–30 ferrets for their job of Rat and Rabbit catchen, a job they would be dewing every year for about six weeks round Christmas Time. Now among them lot a ferrets wus an old female ferret hew wus ever so nasty an would bite anyone excepting Bluff or Father hew tried to pick her up. Even us boors had tew wear a leather glove when we wus rabbiten, if not she would hev us if we went tew pick her up and she wont leave go. Next mornen Bluff wus down at ours, dressed in his only suit, a big heavy gaberdine jacket which had a big inside pocket big enough tew hold a couple of dead rabbits. He wore cord trousers, a chummey hat, a spotted neck kerchief and thick hob nailed boots, and looked what he was, a proper Countryman. Father wus dressed nearly the same, only he hed a clean eck kerchief and wore what he termed his "Half Strained Boots" slightly rger than his every day boots. Now we would say Boots, well they wus allest ‿alled Shews cus there wont sich a thing as Boots in them days, everything was Shews, even if they wus leather and come up tew the knees. Well they started orf tew the Station, walking of course and curled up inside Bluffs big pocket on a bit of Hay was the old Female Ferret and also in the pocket wus Bluffs and Fathers money. I dont suppose they could hev scrapped a pound up between them, but exceptin for a few pence to pay their train fare, the rest wus in a wash leather purse inside the pocket with the old ferret. Crowds of people got off the train at Lynn and nearly all wus maken their way tew the Walks. Arter a little while Bluff wus hungry so each had a Pork Pie then went an had a drink. Each time Bluff pulled the money out they both looked round tew see if anybody wus watchen them but no body seemed tew be a bit interested inem, an they begun tew think their mates had told them a lot of lies about Pickpockets. Then they stopped to watch a bloke who wus selling Gold Watches and Chains, for as little as Five Bob and hew wus even given gold Tie Pins away because he sed he wanted "Tew make everybody happy". The crowd round this bloke started shoven although they couldn't move Bluff far, he wus like a brick wall. They wus proper interested in this bloke an even forgot Pickpockets until there was sich a commotion and a smart bloke next tew Bluff started shrieking an Bluff felt him pulling his hand out of his inside pocket. As his hand come intew view so did the ferret, she wus still hanging on. Bluff nearly went crackers, he snatched the ferret off the blokes fingers, shoved her back intew his pocket, collard the bloke round the neck and shouted "If you had hurt that ferret, Ide twisted yar blasted neck round six times" and knocked the fellars hard hat right over his eyes. A policeman wus soon there and wanted tew know what it wus all about so Bluff said "That dressed up B . . . there tryed tew pinch my best ferret". On being asked if he wanted tew hev him charged Bluff said "No I hev larnt him a lesson what he ont forget in a hurry cuss hus blasted fingers will be tew sore".

OUR OLD CHURCHES AND CHAPELS

Now there is one thing that I reckon is all wrong, that is dewing away with all these beautiful old Charches an Chapels. I was over at my birthplace a week ago an my brother was telling me the Rectory at Magdalen is for sale, he said there ean't any Church services only about once a month an as there ean't a Parson at Magdalen at orl, so somebody come over and preach, theres only about a half dozen people go. Now hew's fault is it that things hev got tew that state? myself I think its acuss they don't make the sermonds interesting enough an don't keep up with the times. Now I arnt a religious sort of bloke, my motto is "If you can't dew anybody no good, don't dew em any harm" an I don't profess to be a Bible thumper but I don't like to see our lovely old Churches and Chapels let go tew rack and ruin or sold orf for storage places. Cor blast when I wus a kid over 70 year ago, there wus services for the little-uns at 10 am on a Sunday and there used tew be nearly a hundred go; then there wus the Church Service at 11 am or another service for kids at 2.30 pm then a service agin in Church at 6.30 pm. The big Church Bells usta ring and besides the big bells there wus also a full set a Hand Bells an a lovely old Organ. There quite a big choir for a small village like Magdalen an altho' I don't understand music an this wus years afore there wus sich things as Wireless I heard a gentleman from Leicester say as he stood listening outside tew the Choir "My word, thass lovely singing, I would give £10 tew hear that music on a gramophone record." I hev allest understood that the Church are the richest people in the country, I don't mean the parsons, but the Church of England, and I put some of the blame on them for letting things get in sich a state I wus born not a underd yards from Magdalen Church also my Father and his Father afore him an I hev heard him tell tales of things that had happened concerning the Church and Vicarage. In them times, the Parson used tew hev a Groom and Gardener and there wus tew Sarvants in the house an if anybody needed help they allest went to the Parson an wus never turned away. My fust recalection is bein in an old three wheeled pram, watchen the burial of Parson Taffy Davies an I wus bein looked arter by my brother Sid. I bet I wornt more than tew or three year old cus there is only about tew year between each of us 13 kids so if I had abin any oder I shont abin in the Pram. I ken still see a Bishop an several Parsons round the grave but sorry tew say our peepin over the Church wall ended in Tears cuss my brother or one of his pals put his foot on the old Pram an ass over hed I went out onto the road. During my life time I hev seen all the lead on the roof taken off and the lead has all bin melted and run onto sand frames in big squares then all put back on agin. The work which was done by Mr Riches of Watlington wus all done with Blocks and Pulleys if it had a bin now no doubt they would a used a huge crane. When I fust remember and wus Moll Catcher on the Ouse Banks there wus Churches and Parsons at Magdalen, St Peters, St Germans and St Marys, now with Magdalen nearly bein on the scrapheap there is only St Germans left. Its the same with Chapels, there wus three Chapels in Magdalen, now there is only one. Fust the Fen Chapel wus closed; it wus a lovely little place, and although it was right down Magdalen Fen it was allest packed on a Sunday and for their Anneversary, they used tew hev a three poled Marquee. Then the lovely old Babtist chapel that wus sold a few year ago, and the last I heard on it it wus a Band Room, but what it is now I don't know cuss I think somebody told me they hen't got a Band now. At Coldham (nr. Wisbech) there wus a Church and Chapel, now the Chapel is sold for a storage place, and there is a service in the Church once a month, the preacher is the Parson from Friday Bridge as the Rectory has bin sold years ago. I think if the Churches and Chapels med their Services a bit more lively and entertaining, adults and children would go as they allest used tew. I know things hev got to change (orfen not for the good) but tew me, it dew seems a pity that those old buildins should stand and crumble away, cuss they musta bin built well cuss some on em has stood there six or seven undered years perhaps longer. Before the Cemetary wus opened in Magdalen in 1900 Father used tew dig graves in the old Church Yard and he has told us of times when he dug onto a grave where they looked as if the corpses hed bin chucked in anyhow, and the Parson hew wus there then told him it wus the Time of the Plague, the Black Death, and in the Church Records it said those hew burried them wus allowed tobacco to smoke to stop them getten the Plague. There wont any sign of coffins so he thought they wus chucked in jest as they died. I know he had a Skull and jaw full of teeth that he brought home and he allest kept them in his old Bureu.

Som'a the preachers in the Babtist Chapel never got paid they jest done it cuss they believed they wus right in what they told the Congregation and I'm sure some on em couldn't read, an theyusta use words that wont in no dictionary but they put their heart and soul intew it. Most a the preachers usta ride old bon-Shakers (Solid Tyres Bikes) from Lynn an one old chap biked right away from Fincham, a distance of 11 or 12 miles and all they got wus a free tea but they wus happy. I must tell yew about one old chap hew usta preach once every six weeks. Him and his old lady rid a "bicycle made for tew". It hed fower solid tyres wheels and they set side by side and they steered by handles more like a box handle on a fower tyned fork. The old bloke wore a cap with tew peaks a double brested cord wesket, and allest hed hs Yorks (leather straps under the knee) on, while the old lady had a handcher tied round her hed, and her dress had anuff material in it tew make a Bell Tent. The old chap, his name wus Tuck, wus a Shepherd, and one arternoon he got tellen on us about I believe it wus *Daniel,* hew wus looken arter his Father's sheep when the Wolf (there might hev bin more than one) but anyhow poor old Tuck got so carried away, he shouted out "Then what did they dew, cor blast, the old Yows (ewes) jumped the Hardles (hurdles) an away they went, cos I bet he hant his dawg with him." Another nice old boy wanted tew get on the Plan (become a regular preacher) so the Minister said it would be alright and his fust date wus at Magdalen. Arter the fust hymn he wus ready tew start Preachen so the young lady hew played the organ set a little bit back, crossed her hands on her lap and looked up at the Preacher but he dint make a murmur, he jest stood there and never opened his mouth. Some one took him a glass of water, cos they thort he had had astroke and told him to set down afore he fell down and somebody suggested they not hev a sermon but jest sing a few hymns. At the finish of the singing, and people ask him what wus the matter he said "I was alright until I looked down from the Pulpit and looked into our Organist's eyes an her face reminded me of Jesus an I was tongue tied." I recon he like the look of her! Then once a year would be Camp Meetin Sunday, it wus held outside in somebodys grass field and a farm wagon would be drawed in ready on the Saturday and seats set all round it. About tew o'clock Sunday arternoon people would begin tew gather round, then they would march up Church Road tew Magdalen Bridge, along the river bank, back tew the wagon, an all the while the people sung and a fellah would walk backwards playing a concertina an the tune wuss all the same "I am on my way to Heaven, singing Glory Halleua" cuss we are not *Planters,* but *Ranters* (Chapel Worshopers wus allest called *Ranters Roarers*). Another thing there usta be a lot on wus when the Preacher wus preachen, someone would shout out "Praise the Lord" or "Halleua", or "Amen" but they wuss all happy and thought a lot more of that than people dew nowadays at Greyhound tracks or watchen Speedway.

WATER

I hev jest bin thinken there is all this slaver in the papers and on T.V. about the water problems jest acuss we heant had any quantity of rain now for some little time. I hev knownit a lot wuss than this and then there wont half the song about the drought as what there is now. As I go about I see thousands of tons of water wasted where years ago there wont the water tew waste. We hed tew manage on rain water for everything, drinking, washing, for cattle and for the land. I reckon there is more water wasted on swimming pools, car washing, at factorys, an hundreds of other things than ever need be. Cor blast now some people eant happy if they dont hev a bath twice a day which I reckon is all squit, cuss when I wus a kid there wont half a dozen bath rooms in the village so people hed tew use the old zinc washing bath, sort of wash one leg at a time and the rest after. Some old people, my parents included, dint hev a proper bath from one year tew the next an I well remember a lady tellen me that her father hed never hed a proper bath since he wuss a baby and he lived to the ripe old age of 97 and right up tew a month afore he died he used to hev a tew mile walk every day. Us kids used to be bathed every Saturday night but we all used the same water and arter we hed hed our bath Father would say "I think I will wash my feet" so jest to freshen the water up a bit Mother would put another kettle of hot water in. Mornings first thing (about 5.30 am) we all hed to wash our face and hands, but all used the same water, dinner and tea time, jest the same, although when we hed all finished washing the water looked a bit like pea soup, we dint ail anything and wus allest healthy and well. There wus no tap water so we hed a cistern and during a period of heavy rains we hed tanks and wooden barrels which we used to fill up and there wont be a drop wasted if we could help it. Sometimes during a dry spell there would be a lot of little squiggly insects in the cistern or there might be a slug or a worm or tew get in but arter the water wus boiled it wus as good as tap water, or to my opinion better, cus rain water med better tea and there wus no skum when you washed. For drinking purposes we had a charcoal filter and arter the water hed bin threw that it wus as clear as gin an dint smell of that old stuff they put in now a days tew purify it. There wus lots of places where there wus lovely cold clear water from some under ground spring, and there wus a kind of drinking place for cattle to drink, or in hot weather the cows and hosses would stand in up tew their knees. One place I allest remember wus at Saddlebow (about 2 miles the St German side of Lynn) right in front of the Bull public house. It had a gravel bottom with a nice sloop tew it and beautiful clear running water and it was nothing to see the drivers of hosses and carts go inside the Bull, and the hosses go over the other side of the road to the shallow pond. Nearly all those watering places hev bin filled up so that water is still draining away somewhere jest like the Ponds that wus in every grass field, some on em wus never dry however long we hed a drought or if by chance they did, a hole would be dug down intew the clay bottom and it would soon fill up with water. It might be a bit black but the stock would drink it and it dint kill em. Another thing I heant seen for years are the Dew Ponds; in dry times a shallow hole would be dug and there wus soon some nice clear water in it, but I dont suppose there are any old timers left hew knew where to dig a dew pond. Its like everything else as soon as there is a quicker and easier way of gitten anything they dew away with the old method but it dont allest pay, then as soon as we git a dry spell they are in a rare old muddle. Its jest the same with the dykes in the Fens round here, they keep making them deeper and draining the land so that the water is 6 to 8 feet from the top then they moan acuss there eant any moisture in the land. One big land owner hew lived not far from me once told me that he liked to see water in his dykes *all* the year round. I wus thinken about that only yisterday when I wus gassing rabbits on a big estate. Some dykes hev bin deepened so much I hev to hev a 20ft ladder tew git down to the rabbit holes in the bank. They are so deep that tew slip down the bank would mean you hev got to stop there til somebody come and pull you out with a rope. You wont drown cuss the dykes hev bin drained so much there eant 6 inches of water in em. Of course now they hev irrigation sets pumping water miles from a river or drain, it does good but where the huge sprayer stand and where the tractors hev tew run about on the cropping arter its bin soaked, the old strong land set like concrete. But I expect its all good for trade.

Another thing that uses up millions of gallons of water very every day is flush toilets where years ago there wont sich a thing. Every time the chain is pulled it mean another tew gallon of water is used, where years ago the old Vault Privy stuck down the bottom of the garden, would last a year without being emptied. I know the flush toilet is cleaner and much handier but very often they are used unnecessary and its water wasted. I know what it is tew drink water straight out of the river Ouse at Magdalen. We never had tap water until I wus about 8 or 9 years old, then when I moved from Magdalen to where I live now in 1925 there wasn't tap water here so it wus back to soft (rain) water again. To keep the water pure and to kill any insects etc. in the cistern a live eel would be kept in, but once at Magdalen Father put a few lime cobbles in. It did kill the insects etc., but when Mother done her washing it nearly skinned her hands and arms so that put paid to that. I dew think people are much tew fussy about what they drink, I eant sayen if there are stinken dead rats in the water that dew anybody a lot a good but I hev drunk water from various places an hev never caught any complaint yit.

I remember once being with Father mole catchen and blast it wus a hot day after a very heavy thunderstorm. I wuss dry, and in the grass field where we wus setten some traps wus a low full of nice clear water, left from the very heavy rain the night before. I kneeled down on some mole sticks and hed a good drink and never tasted any thing so nice in my life and 65 year arter I am still alive tew talk about it.

Yes, on the farms, thousands of gallons of tap water are used every day for mixing with spraying stuffs for weed killin (never see any hand hoeing done now) for washing pea vinners out and peas before they are canned in big factorys all over the country for various jobs, in fact there are thousands of jobs where water is wasted, well thats my idea.

I expect you think this is a rum subject, but it often cross my mind when I see water wasted its cos people wont old enough to remember the times when we hed tew be sparing with water, like arter a very dry spell when Mother would say "I cant hardly dip a bucket in the cistern, so hope we soon git some rain". Perhaps the hole dug in the bottom of the drain opposite our house down at Foldgate Lane wus nearly dry so that meant the hole hed to be dug deeper and then the only way to get the water wus to tie a rope on the bucket, stand on the plank over the hole and drop the bucket in and pull it up hand over hand. I ought to add that the water from the hole was for the horse, pigs and chickens, for our household use, when the cistern and tanks wus empty we put a couple of tubs in the cart filled them up in the village put a clean piece of plank in the top of each as floaters and tied a clean piece of sacking round the tops to stop spilling so much. We dont want those times back again but in one way I think it would make people think twice about wasting water.

SOME RARE OLD COOKEN RECIPES

I wus thinken a few weeks ago what a lot of rum people there is in this World, mind ya I heant bin out of England but by the newspapers and television it seem thers rum people tew be found all over. Mind yew besides a lot of badons there are a lot of gooduns, an I believe I know some a best on em.

There is one very old couple hew I know an if every one wus as good as them this world would be a lot better place than it is now. Both on em are older then me an as I am 75 you know they must be gitten up the tooth a bit but as I say they are lovely old people and wont hurt a worm. I call an see em about once a fortnite when I am round their way an the fust words they say is "Your got time for a cup of corfee, heant yar" an I will say this The old lady make the best cup a corfee I ever tasted, an she allest git some of her cakes or mince pies out an she ken cook an all. Well we got talken about cooken and she happened tew say she wus hed cook at a Titled Gentlemens house when she wus 18. She told me every mornen the Lady of the house would come down and tell her what tew cook and how many there would be tew lunch etc., then arter that she say "It was allest left to me". I asked her what sort of stuff she cooked an as she told me I writ some on em down, so I'll tell yer jest as she told me although yew mussent blame me for mistakes cus I praps dint catch the names right as the old lady is a rare quiet talker, but anyhow here is a few what she told me.

Oxtail Soup
Tew oxtails, a gallon of water, 1 oz salt, 4 nice carrots, 3 decent sized onions, 2 turnips, 1 hed of salary, 6 chives, $\frac{1}{2}$ teaspoonful a peppercorns, some herbs

an a bit a ham. Soak an wash the tails, pour on them the gallon a water, bring tew a boil, add salt and dont forget to clear the scum as it form on top. When the scum stop rising add the carrots, onions, etc. an stew gently for 4 hours. Then lift out an divide the tails at the joints, an add arrow root. A little wine added will help.

I bet a basin full of that would warm the cocckles of yar heart on a cold day. Another one she told me wus:

Rook Pie
6 young rooks, $\frac{1}{2}$ lb rump steak, $\frac{1}{4}$ lb butter, $\frac{1}{2}$ pint stock, salt and pepper. Draw and skin the birds, remove necks an backs an split birds down the breasts. Put birds in a deep pie dish, cover each bird with them strips of steak, season well with salt and pepper an add small pieces of butter an add as much steak as will three quarters fill the dish. Cover with pastry, the fust $\frac{1}{2}$ hour in a hot oven tew make the pastry rise, them more slowly for another hour tew allow the birds tew be properly cooked. Arter they hev been cooken an hour, brush the pie over with yolk of egg tew give the crust a shine.

Norfolk Dumplings (Twenty minute swimmers)
All round here where I live, in fact all Fen People, call Norfolk Dumplings or Light Dumplins "Twenty Minute Swimmers" and you will see the reason why when I tell yew. First you put $\frac{1}{2}$ lb plain flour intew a pudden basen with $\frac{1}{2}$ teaspoonful of baken powder an mix together with $\frac{1}{4}$ pint of cold water. Divide mixture intew 4 balls, then drop them intew boilen water an watch out an dont let em stick tew the bottom of the saucepan. Let em boil for 20 minutes. My Mother used tew put hers intew a big boiler which had $\frac{1}{2}$ bullocks hed, onions, carrots etc. in it and which hed bin boiling for about $1\frac{1}{2}$ hours, or until the meat boiled off the bones. As soon as the dumplings hed bin boilen 20 minutes they are ready tew serve an if you heant ever hed any now is the time tew start cus they and the meat are lovely.

The reason Mother med about a dozen and a half dumplings wus there wus 13 in our family and we could'nt harf clear some grub up, but if you got a belly full of meat, dumplins, potatoes and carrots you werent hungry for an hour or tew.

Dumplins could also be med with suet or yeast and often if Mother hed bin baken bread she would save some of the dough and roll up intew dumplings.

Queen of Puddings
1 pint of milk, $\frac{1}{2}$ lb breadcrumbs, $\frac{1}{2}$ tea cup of white sugar, yolk of tew eggs, rind of lemon grated and a bit of butter about $\frac{1}{2}$ size of an egg.

Pour the milk (hot) on the breadcrumbs, then stir the sugar and butter (melted) intew them. Beat yolks of egg add lemon peel an a little nutmeg. Bake mixture in a buttered dish till its done.

Beat whites of eggs to a froth an mix with $\frac{1}{2}$ cup of white sugar and add juice of a lemon. Spread a layer of Red Currant jelly over the pudden. Pour whites of eggs over the whole, and bake tew a light brown.

Arthur's kitchen.

Now I will give ya my fathers favorite pie.

Pluck Pie
Years ago anyone could buy from the butchers the hul inside of a pig such as the heart, liver, kidneys, lungs etc., jest as they wus took out of the pig. The liver and Fry mother would either fry or make intew a pie and dumplins but the Pluck (or Lungs) she would make intew a special pie for Father. She would cut the Pluck intew smalla bits, add currants, chopped apple and sometimes pumpkin, put intew a dish with a covering of dough all over the top and bake in a moderate oven. Of course she would put an egg cup upside down in the centre of the pie dish, just tew keep the crust from fallen in.

Other Dumplins
Apple dumplins, meat dumplins, currant and jam rolls would all be med with the same class of dough. Just the dough would be rolled out flat, layed on a greased pudden clorth. The apples hed already bin cut up intew little bits and the bits wus then heaped intew the middle of the clorth, the ends brought together over the top and tied an it wus all ready tew put intew the saucepan and boil. The beef for the meat pie had already bin boiled and that would be placed in the pudden clorth (sometimes in a big basin, with a crust over the top) and boiled in the same way. Currant or jam rolls would be med by rollen the dough out flat, the currants or jam spread out over the dough then rolled over and over until it med a roll nearly a foot long. The clorth would be tied round in 2 or 3 places and boiled in a big saucepan. Brown sugar (sometimes called corse sugar) wus allest used tew sweeten apple dumplings. Of course in them days the amount of meat cooken depended on what money Mother had tew spend and very orften a meat and onion roll or dumplin would be called "A Hundred Tew One Roll" which in Norfolk meant "A Hundred bits of onion tew one bit of meat".

Bread Making
I hev already mentioned about our large family and how Mother usta bake all her own bread (8 big loaves, 3 times a week) so now I will tell ya how she usta dew it. The evening before she baked she would put the right quantity of Plain Flour intew an earthenware pan (called a Panshon) mix in the yeast either Brewers Yeast *wet* or Dry Yeast called German Yeast. Add a little salt, a few *cooked* potatoes mashed up then add the water, slightly warmed.

She would knead it for several minutes until it wus jest right, then cover it over with a white cloth and when we went tew bed it would be stood near the fire tew rise. Next mornen when she had got our brick oven nice and hot she would grease the bread tins and put jest the right quantity of dough in each tin and bake until the loaves wus nice and crusty. As the brick oven only held 4 tins at once she hed tew hev two lots of bread baken, then while the oven wus hot she would cook some pasty or anything else she wanted tew cook. Mother allest brought her flour 10 stone at a time and kept it in a wooden bin in the kitchen.

Pigs Belly or Chitlings (Hogs Guts)

To finish this off, I must tell yar about how Mother used tew cook Pigs Bellys or Chitlings. We lived quite close tew a butchers and Wednesdays used tew be the killen day. The butcher would send 3 or 4 sets of bellys down just as they wus took out of the pigs. Fust thing she would pick a lot of fat off the bellys which she would Render Down intew lard. She would then deal with the *contents* of the bellys (what a stink) turn them inside out, scrape them well, run them through agin with clean water then leave them all in salt water, till the next mornen. Thursday, Friday and Saturday mornens she would turn them inside out agin then put the big old Iron Boiler over the fire and cook em for about three hours. They wus then ready tew eat and what wus left would be fried up tew make us all another good meal. The fower sets of Bellys would only corst two bob (6d a set) and if any of yew hant ever tried "Fried Hogs Guts" yew dont know what yew hev missed.

Some people would turn their snout up or pretend tew to be sick if you mentioned eaten hedgehog but if you hev a mind tew believe me its very good grub. If you had it cooked like we usta hev it you wont know the difference between hedgehog and rabbit only if anything hedgehog is the richest and the fat is a bit more yellar. We dint eat ours like the gypsies used tew hev theirs which is cooked with the jacket on but Father skimmed them and they looked jest like rabbit. The fust one I tasted was cleaned and cooked by an old bloke Billy Singer (always called Billy-ma-dear). He lived in a van which today would be worth about 2 or 3 hundred quid an he asked us boors if we wanted tew try it. I had a bit an arter that we often had one and the prickly skins we tacked up in the stable tew stop our hoss from rubbing.

ODS AN INDS

About six months ago a smart old gentleman came round tew my door an asked me if I knowed him. I looked hard at him then said, "No Boor, that I don't". He said, "Your will dew when I hev told your hew I am. I'm Tup Trollope what you went tew school with at Magdalen" I said "Cor Blast, I shount hev known yar, Com on in" so he say "I got my son outside in the car".

I fetched him in and he wus a smart fella about 40. When I knew Tup he wus a raggety Arsed Boor an I dont think he ever washed except when we went bathing in the river. Coo we got talken about olden times and one bit wus when we tied a dead rat on the Chapel Door. I'll tell about that little bit, theres nowt in it rearly but at the time we thowt it rare fun. It wus a beautiful moonlight night an we wus kicken about in a hedge bottom an put a rat out, coo it wus a wopper an as it run passed me I give it a proper broadside with my foot an killed it. Arter we had kicked it about a few times one of our gang said "Lets tie it on the Chapel Door". They wus busy singen inside so we tied it on the handle, then give a knock. Poor old Pint Roberson com out folled by poor old Mrs Bates an when she saw the rat she screamed out "Oh my God father, fetch the policeman Im goen tew faint". Now it wus during the 1914–1918 war and our policeman had been called up an the St

Germans policeman wus dewing both villages an we thought there wus no policeman round us for miles, but we got a shock cus at that very moment hew should come up but P.C. Jackson (the St Germans Bobby) we wus scared althow we hant no need tew be cuse he wus a man nearly 60, an looken back he couldn't run half as fast as us, we darted all ways an runned like hares. Me and Tup runned down the Mill Road an got down the drain behind some thorn bushes an darent move for over an hour but at last we ventured out, an at last met the rest of the gang an herd that the bobby had got on his bike an wus half way to St Germans by then. Blarst I wornt harf frightened.

Then Tup and I got talken about another school pal – "Stallion" Wing. Poor old boor, he hant no mother an his father wus a cruel old sod, allest knocking him about but at heart he wus a good old boor (Stallion I mean). One day he got on old school master Carter's nerves so much that he shut him up in the coal shed. When he wus let out he had amused himself inside by breaking all the lumps of coal up intew dust – coo old Carter wus savage but Stallion dint care cus I allest reckon he had bin knocked about so much, at home and school, that they had knocked one devil out and two in. Then there wus the time when Stallion wus up the Parson's apple trees and the Parson, Rev. Joase, caught him. He asked Stallion what he wus dewing and Stallion said "I hev got tew take a spider tew school tomorrow and I thought I might find one up here". When the Rev. saw the apples in Stallions jacket pocket he said "An what are the apples for?" so Stallion said "Thas tew feed him on tew night". Tups father bought some shallots off my old Godfather Bluff for about a bob or eighteeen pence and dint pay him. Old Jack kept asking for the money an Bluff would say "Dont your keep worry gutten about that, I shell pay you one day". At last old Jack said "You keep sayin one day, jest tell me what day" so Bluff stood for a minute and sed "Now judgement day look like being a busy day, so Ile pay you the day arter". Coo it wus nice to hev a jaw about when we wus boors, an Tup said "I should hev bin over here afore only I dint know where your wus or if yew wus alive then blast if my boor browt a book home that your had writ", and then his son chimmed in and sed "Theres bin no rest until I browt him over tew see you". I set arter they had gone an thowt about lots of things on the rum old boors and mawthers that there wus at Magdalen 65–70 year ago. Times wus hard but we still had our fun only in them days we med it ourselves. All kids except the Upper Tens wore thick leather boots (allest called shews) with big hob nails in the soles, its a wonder we could run but we could and played "Nany All" or "Burglers and Bobbys" in the old school yard or "Shewen Horses" like we hed seen old Bill Lambert dewing up the Blacksmiths shop. Coo tew see that shop full of horses and perhaps a half dozen waiten outside, some on em wanten a full set on (4) med you wonder how old Bill an his mate would get em all done. Young hosses that hant bin shod afore had tew go threw it. The fust time they kicked out or put all their weight on Bill he would git his big twitch down (a big strong stick what had a hole bored threw at one end and a strong bit of cord threw it) and he would give the hoss a few right up the ribs an keep shouten out "Whow Collabar". Sometimes a young hoss had

what wus called a Wolfs Tooth, which wus a sharp tooth that cut intew its tongue and then the hoss dint eat properly. Bill would put the twitch on its top lip, give the twitch a few turns, then git his chisel and tap the tooth in the exact place to make it break orf. I hev thowt when I hev bin heving a tooth out at the Dentist that that is bad enough but not half as bad as them young hosses went threw.

SPYS

The 1914 War had jest started and everybody had got War Fever. It wus a long while since there hed bin a War so close tew us here in England and all sorts of yarns got about that the Germans wus landen boat loads of Men and Womin in this country to spy out what we wus dewing. Father and Mother, like thousands of others, wont no scholars and believed all what they was told. Anybody fresh in the village wus soon under suspicion and wus watched by the locals. A bit later when the Zeppelins com over and dropped bombs on Lynn and other parts of East Anglia there wus soon tales floaten round that somebody had seen somebody signalling to them by motor head-lights etc., and that med people keener than ever. I think half the people expected tew wake up one mornen and find the Germans walking about up Church Road. I know one mornen I wos with Father molecatchen and we met a Bill Barrett hew wos foreman for Mr. Peacock. He got tellen us that anyday he expected tew see the Germans marching along the High Road

(Wisbech Road) and he sed he should say "March on me boys and blow the Big House up", meanen the House of Parliament. Father said "Dew your mean it" and he sid "Yes Teddy, thats jest what they will dew". Arter we left him thats all Father kept on about and I believe if we had seen a gang of people comin towards us we should hev hid up, or perhaps done somethin much wuss. About an hour later we wos at St. Marys and met a man named Charley Self, so corse Father started tellen him what we hed heard. Charley larfed and sed "No Ted, they ont git as far as this, we got a Navy between them and us so they will hev a job tew march in here". With that bit of news we both felt better, else afore that neither onus dint want our lunch. But there wus still the Spy situation. We lived about a mile out of the Village, down what wus called Foldgate Lane, down the Fen, and at that time there wus only our house down there. Our house wus set in the middle of an orchard, with a road-way runnen past the top end and the other three sides wus surrounded by grass fields. One mornen my old Godfather Bluff had come down to help Father tew cut a high thorn hedge down and had brought his slash hook with him. While Father an Bluff wus cutten the hedge down, I wus busy cutten the big bits out to saw up inter logs to burn on our house fire. I looked up and saw Mother beckenen tew me and she hed her finger tew her lips. I went tew see what she wanted; and she jest went "Shus" and pointed threw the thick hedge to a smart looking fella that wus setting on a stool with a lot of papers in front of him. He looked tew be drawen suffen and Mother said "I bet his is a German Spy". She sent me back tew tell Father and Bluff while she watched the bloke. Us three went and hed a peep an Father said "He's a Spy right enough but he aint goen tew git away from here alive". Father and Bluff whispered together then they med up their mind what to dew. I wus sent tew git our old Highwaymans Pistol, Father went arter his old muzzle loader and Bluff wus tew use the long handled Slash Hook. Up tew then I hed never bin allowed to fire only gunpowder but this mornen Father put a charge of shot in, had a look tew see if I had the cap on firm and I wus fixed up. He loaded both his barrels, then I wus ordered tew go with Mother up tew the house and "If that Spy run your way I wus tew shoot him at the top of his legs". Father and Bluff then went threw the hedge and when they got close tew the stranger Father walked tew about 5 or 6 yards in front of him and Bluff about a yard behind him. The man looked up and said "Good Mornen" then Father asked him "If he wus a German 'Spy'", so he replied "No I am English" to which Father replied "Thas good job for you cuss if your had a bin a spy, I shoulda shot you" and Bluff said "Yes Mate an if Ted hada missed I shoulda had yar hed orf". When the stranger had explained hew he wus everybody was happy. He wus from London and wus staying with his Uncle Banker Sharpe and wus dewen a drawen of tew cows and calves and wus maken a very good job onem. Although at fust he wus scared and when Father asked him in tew hev a glass of Elderberry Wine everybody wus all larfen but I noticed every now an agin he kept lookin round at that Bill Hook that Bluff had stood agin the door.

This next Spy scare affected everybody in Magdalen, in fact everybody for miles round. Magdalen, where we lived lay beside the Great Ouse river and the village itself wus several feet below high water level, so there wus only the Banks to stop all the Fens from being flooded. At high tide the masts of the Fishing boats showed above the Banks. Tew keep the Fens drained a pumping engine would pump the water from the Fens intew a channel and that would empty itself intew the Ouse threw a sluice at low tide. If anything had stopped the sluice doors from shutting properly the water in the River would run back and would soon hev flooded all marshland. Our Drainage Committee had a meetin and with all the Spy Scare decided to have an *armed man* to guard the sluice during the hours of darkness. Patten Plaice wus the man picked for the job. Now I must add that Patten wus Bluff's brother, and like him; an my Father, wus no scholar (dint know a Big A from a Bulls Foot), but he wus very keen on his new job an I believe the old boor sort of thought he wus important, sorta like some high ranking officer in the Army. Every night about 7.00 he would be seen walken along the river bank from his home towards the Lodes Head Sluice, and he allest carried his double barreled gun on his shoulder, and walked 3 inches more than upright. Well everything went alright he liked his job and he had a place where he could get in out of the rain etc., which Charley Benefer, a local fisherman, let him use. Now this place wus jest level with the top of the river bank, it wus a sort of store place where Charley used to store and mend his nets, the bottom room wus 10 foot below the bank level. It wus one Sunday night or early Monday mornen when Patten got in disgrace. He hed gone in his shelter tew heve his bit of bread an cheese and everywhere being so quiet he dropped off tew sleep. He wok up with a start, somebody wus going past and when Patten saw somebody aginst the sluice, his mind acted very fast, it must be a spy tryen tew "Blow the sluice up". He grabbed his gun, ran as fast as he could down the hill to the Lodes Head Pub and wus called "Stout, Stout, there's a Spy!" Stout Pilgrim, the landlord jumped out of bed, slipped his trousers and slippers on, an rushed downstairs. Patten wus out of breath but managed tew tell Stout that there wus a spy and he tried to blow the sluice up but he tackled him and wus now racing up the bank. Stout grabbed Pattens gun, and chased off arter the spy, but went in the wrong direction. Stout was much nimbler than Patten who wus what we called *Splod Footed* and took up nearly as much room when he walked as a Cart load of muck. Years ago there wus fences about every mile or so along the river bank which stopped cattle from straying, so when Stout got to the fust one he tried tew jump it but caught his toe on the top plank and Asse over Heels he went and the gun went off. The report of the gun woke half the people up in Magdalen and people thought the Germans had arrived. Falling over the style Stout badly sprained his ankle. People got up and called the village policeman who got a statement from Patten, so he got on his bike and dashed off down tew Magdalen Road Station where he found the man setten in the Waiting Room. When everything hed sorted itself out, the man hew Patten saw, wus a man hew had bin over tew his brothers down Magdalen Fen, an as he wus a postman on duty at 5.00 am on Monday mornern at Lynn, had walked along the river bank cuss it wus a short cut, and wus catchen the Mail Train from Magdalen Road to Lynn. There is no more tew say except that as we should say nowadays, Patten wus med Redundant, and that finished "Guarden the Lodes Hed Sluice". Stanley Riches the blacksmith med Stout an iron cross for his breavrey! Another spy tale which is a bit different wus when a man I knew very well gave another man a lift from Littleport tew Ely. The man I knew wus an agent for J. Whetstone, a firm that bought an sold thousands of tons of Taters every year. The agent wus on his way tew Ely, saw a man walking and offered his a lift. He accepted a ride and although there wont a lot of conversation passed between them, something what the stranger said med the agent suspicious. He asked the stranger where he wanted droppin an he sed the Station, but instead of takin him tew the Railway Station he took him tew the Police station that turned out he wus a "Real German Spy" or so a man I know told me.

Dick Joice and Arthur Randell.